PRIMORDIAL ALCHEMY
& MODERN RELIGION

Essays on Traditional Cosmology

Rodney Blackhirst

PRIMORDIAL ALCHEMY & MODERN RELIGION

Essays on
Traditional Cosmology

SOPHIA PERENNIS

SAN RAFAEL, CA

First published in the USA
by Sophia Perennis
© Rodney Blackhirst 2008

Series editor: James R. Wetmore

For information, address:
Sophia Perennis, P.O. Box 151011
San Rafael, CA 94915
sophiaperennis.com

Library of Congress Cataloging-in-Publication Data

Blackhirst, R. (Rodney)
Primordial alchemy & modern religion: essays on
traditional cosmology / R. Blackhirst. —1st ed.

p. cm.
Includes bibliographical references.
ISBN 978-1-59731-083-3 (pbk: alk. paper)
1. Cosmology. 2. Alchemy. I. Title. II. Title:
Primordial alchemy and modern religion.
BD 511.B57   2008
110—dc22   2008005659

CONTENTS

*In the man who has realized Adamic perfection,
everything is original; his being is fully awakened
and united with its origin.*

Titus Burckhardt

Introduction

THE writings brought together in this collection dwell upon a primordial cosmological symbolism that is well described as *alchemical* because it has persisted in the various alchemical traditions in a particularly complete and direct way, but it might be more precisely described as *aurumic* because of the central place of the symbol of gold within it, and because it is not by any means confined to the alchemical traditions, formally speaking, these being only one of its manifestations. At the core of this symbolism is the doctrine of the cosmic cycle of the Four Ages and in particular the myth of the Golden Age and its Golden Race. Properly understood, alchemy refers back to this mythology at every turn and all alchemical symbols must be understood within such a context, for the Great Work of the alchemist pursues nothing other than the 'golden thread' that runs from one Golden Age to the next, through all the declining Ages of Silver, Bronze, and Iron, and concerns nurturing and preparing the way for the future Golden Race whose seeds are laid now at the close of the current cycle. Astrology, which is a sister science to alchemy, is also primarily concerned with cosmic cycles and cyclic regeneration, and again the notion of the 'Golden Race' is the key to understanding its integral symbolism. The various symbolisms and subjects explored in these essays have this doctrine underpinning them, often in its Biblical form of the myth of Eden and the earthborn Adam. In Islam, in keeping with that religion's inherent primordiality, we find the Adamic autochthony expressed in the axial symbolism of the canonical prayer in an especially illuminating way. Most of the matters covered here concern either a primordial threefold symbolism (Sun, Moon, Earth, for example) or a fourfold symbolism (north, south, east, west) that is also an expression of axial duality (horizontal, vertical).

It hardly needs to be said that modern, industrial man—even where he retains a religious affiliation—has lost sight of this symbolism and does not even suspect how relevant to his plight might be the primordial teachings of the broader alchemical tradition. Of all the traditional sciences it is alchemy—based as it is in metallurgy—that is directly concerned with the coming of the industrial order. In alchemical terms modern man lives in the Ferric Age and his state is best analogized to the properties of the metal iron—hard, cold, unbending but quick to succumb to corrosion and rust. The great ancient wisdom traditions of the world all anticipated this present age for it was already implicit in the technological and other changes that brought on the dawn of history. These ancient traditions, dismissed as childish superstitions by the scientist, contain the keys to self-understanding that contemporary man so desperately needs. It is no exaggeration to say that in recent decades the sense of crisis in the modern world has become overwhelming. The toxic fruits of the 'reign of quantity' are everywhere to be seen. Technocrats and scientists stumble for solutions using the same limited modes of understanding man and the *kosmos* that precipitated the crisis in the first place. It is increasingly clear that humanity took a wrong turn at the outset of the industrial era and that we need to look back over our mistakes and attempt to recover something of the wisdom that we so recklessly cast aside in the insane rush to conquer nature. Alchemy, astrology and the hermetic sciences no doubt became crass superstitions in popular hands, but hidden within them is a timeless *aurumic* core, a profound philosophy of the *kosmos* that addresses the realities of the human condition. The essays in this collection are intended to place some of the keys to this philosophy within the reach of contemporary readers.

The other chief source employed in these essays is the ancient Greek tradition which is again approached as an integral wisdom tradition and not, as classicists would have it, a foreshadowing of modern thinking. In the Pythagorean and Platonic strands of Greek thought, in particular, we find a cogent and comprehensive expression of the doctrine of cosmic cycles. The central text in this respect is Plato's *Timaeus* which has served as a cornerstone of both alchemical and astrological inspiration for over 2,000 years. And behind the *Timaeus* is a further source, namely the *religious tradition* of the Greeks and especially that of Athens and Attica and the very ancient cult of the Acropolis. Here we find a religion and accompanying mythology in which the divine smith god fathers a solar and golden child directly from the Earth. It has rarely been acknowledged that this cultus of the Smith God concerning the birth of the golden autochthons, the primeval Athenians, is resoundingly alchemical (*aurumic*) in its symbols and themes, nor is it appreciated that this religion—Plato's own religion—forms the background to the *Timaeus*. In these essays the Athenian mythology—and its autochthony cult which is an expression of the doctrine of cosmic cycles—is understood as the background to Plato's cosmology and ultimately the source of integral alchemical and astrological symbolic orders in the West. It is necessary to explore the ancient roots of hermetic cosmology in Plato in order to revivify and recontextualize the alchemical cosmology that modern, industrial man has lost.

PART ONE

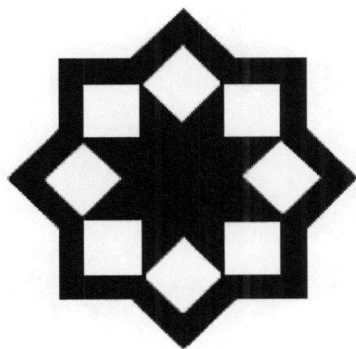

The Mythological and Ritualistic Background of Plato's *Timaeus*

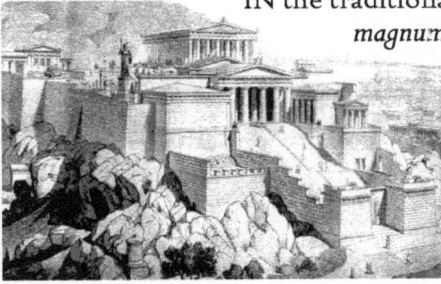

IN the traditional canon of the West Plato's *magnum opus* on cosmology, *Timaeus*, stands in a pre-eminent place. The fact that modern scholars attribute to this work 'more intellectual evil than any text except the *Revelation of John*' is an indication of its significance to traditional ideas in Western civilization; it was pivotal to the premodern world-view. In medieval Christendom it was received as an inspired commentary on the Book of Genesis and regarded as all but scripture. In profane terms, it has the distinction of having been in continuous publication for nearly 2,400 years and the wealth of its ideas still manages to excite some scientific minds beyond merely historical interest. By reputation it is a deeply esoteric text, placed by the Neoplatonists at the center of the Platonic corpus and regarded by them as the key to the divine Plato. The whole Platonic tradition has been nurtured on its teachings. The purpose of this present exposition is to introduce readers to aspects of this seminal text that usually escape attention and yet are vital for appreciating the depth of the wellsprings from which Plato has drawn his doctrines. Modern commentaries are little use in this regard. Typically, they situate the Platonic cosmology as a 'reaction' to the trend-setting Ionian *physiologoi* who are routinely championed as ancient forerunners to modern science. What readers who seek the traditional Plato require, on the other hand, is an understanding based on the traditional spiritual background of his times. It is a singular mistake to import a secular sensibility into Plato and to divorce him from the spiritual atmosphere of his day. Rather Plato's writings—in which, according to the famous Seventh Letter, he never revealed his *full* doctrine, it should always be remembered—need to be considered in the context of the living Greek spiritual tradition in which Plato was born and in which he participated throughout his life. This is nowhere more evident than in the *Timaeus* which otherwise remains opaque and mysterious in many parts.

The central agent of the Platonic cosmology is the craftsman creator god, the Demiurge. Of this character the modern commentators have little that is penetrating to say. A type of orthodoxy was established by Solmsen who argued that the Demiurge is a conception unique to Plato and, in the final analysis, one of Plato's most stunningly original contributions to the history of ideas. But when we remember that Plato was not, as a lot of twentieth-century scholarship seemed to think, an Oxford don, or a 'speculative philosopher' in the modern mold, but a pious Athenian pagan of noble birth, the source of his conception of the Demiurge becomes obvious: it has been modeled on the mythological figure of Hephaistus, the smith/craftsman god of the Athenian religion whose cultus was celebrated on the Acropolis. The identification is most evident in passages where the Platonic craftsman god 'forges' aspects of the *kosmos* in his 'workshop' and where Plato resorts to the language of metallurgy to describe his Demiurge's activities. The long monologue on creation by Timaeus of Locri is replete with the language of crafts, but above all of those of the crafts of fire, Hephaistus being the fire god among Olympians. The religious and mythological background to the *Timaeus* is signaled in the early sections of the dialogue and in its companion dialogue, the *Critias*. Hephaistus is mentioned by name, along with Athena and Gaia. These are the tutelary deities of Athens whose rites and mysteries were celebrated on the Acropolis and in the broader cult of the Attic region. As an Athenian, a son of the Attic soil, they are Plato's own gods. Solmsen and others only ever consider Plato among philosophers, and among philosophers they can find no forerunner to Plato's Demiurge. It is only when we appreciate Plato in his full native context that the model becomes obvious. This in turn illuminates the mythology of Hephaistus. The Platonic Demiurge recasts the Olympian fire god so that we must understand his mythology in cosmogonic and cosmological terms. Thus his crafting of the extraordinary Shield of Achilles in Homer's *Iliad* must be understood as an act of cosmogony, and his binding together of Aphrodite and Ares, Love and Strife, in the *Odyssey*, must similarly be seen as a cosmological symbolism. In over 2,000 years of voluminous commentary on the *Timaeus* only the Neoplatonists seem to have appreciated this.

But the specific aspect of Hephaestean mythology upon which Plato draws in the *Timaeus* is the Athenian/Attic cultus which the god shares with Athena and Gaia. The setting of the work is a visitation to Athens by an illustrious Pythagorean from Southern Italy on the Panathenaea, the great festival of Athena, the most sacred festival in the Athenian calendar and, consequently, the occasion of the most hallowed rites in the temples of the Acropolis. This is the real key to understanding the *Timaeus*: its full context is that, as the Pythagorean speaks, giving his account of creation, the Panatheanea, great New Year and birth festival of the Athenians, along with its mysterious rituals, is being celebrated throughout the city and in particular in the Temples of the Acropolis. The intent of the work then becomes clear. It is a philosophical—or more explicitly, Pythagorean—exposition of the inner dimensions of the Attic religion, its secrets and mysteries.

It becomes plain that Plato has brought his Pythagorean speaker to Athens to reveal certain inner aspects of the festival on which he discourses to Socrates and others. From what else we know of Plato's life—he was the greatest mind of a disgruntled generation of aristocrats who went into self-exile after the execution of Socrates—it is fair to assume that he was unhappy about the spiritual and civil decline of his times and so imports a

Pythagorean to newly inform the Athenians of the deeper meanings of their own feasts and rites. But it would be wrong to regard Plato as in any way a radical reformer. His purpose is to rectify. According to traditional accounts, during his self-exile he visited Egypt, Southern Italy and other strongholds of traditional wisdom. In the *Timaeus* he brings these old contacts to Athens on the occasion of the cities' most sacred rites. The portrait of the Demiurge is not in any way a bold new innovation in the history of ideas: it is a development—in the mode of the philosopher-poets—of the Hephaestean mythologem in its fullness—a drawing out, not a new departure. Plato proposes returning to the roots of the Athenian religion in order to revivify it in the intellectual idiom of his day. It is important to recall that, despite the drama of the trial and death of Socrates, all the evidence tells us that Plato remained a patriotic, pious Athenian to the end of his life, and it is well to remember that his Academy had no resemblance to a modern, secular university, but was a registered *thiasos*, officially integrated into the state religion of the city, was situated in a olive grove sacred to Athena herself and conducted rites and sacrifices according to tradition and law. The Platonic enterprise was to articulate the Greek tradition in and to the times, a deeply traditional and conservative endeavor, not to subvert or overthrow it with innovations.

The particular mythology and religious cult that Plato wants his visiting Pythagorean to illuminate with a renewed profundity is the autochthony cult to which the Athenians and the people of Attica laid claim. The Athenian boast was that they, of all Greeks, were children of the Attic soil, natives, aboriginal in the fullest sense, born from the soil of the land just like the plants and the trees. In Plato's time, in the context of Athenian mercantile colonialism, this boast carried the extra implication that other Greeks should submit to Athenian authority and power. That Plato has a Magna Grecian from 'well-governed Locri' in Athens to discourse on the true or inner meanings of the Athenian's own festival is, then, counter-colonial. Plato is concerned to refill the forms of the Athenian religion with understandings from the periphery of the Greek world, in contrast to the exploitation and subsequent emptying of those forms into a purely exoteric, imperialist ideology. Athena and Poseidon (the other deity whose rites were celebrated on the Acropolis) were Air and Water deities respectively, and had become emblematic of Athenian naval might, the wind and waves that carry Athenian merchandise afar. During the procession that accompanied the Panathenaea the citizens of Athens hauled a ship draped in a woven *peplos* as a sail around the streets of the city and up the steep slopes of the Acropolis to the Temples of the cities' gods. In Plato's Athens, this represented the twin powers of Athena and Poseidon, the cloth of the weaving goddess, the ship of the sea god, who championed the cities' empire through unmatched naval strength. Timaeus of Western Locri, a small Greek outpost on the foot of the Italian peninsula, is in Athens on the Panathenea to tell his hosts that the rites and symbols of the festival are about something far more profound. In particular, the boast of autochthony has a deeper, spiritual and cosmological dimension beyond being merely a device of ethnic chauvinism. Autochthony is the subject of the most sacrosanct secrets of the Acropolis cult and it is Plato's purpose to restore understandings of its deepest and therefore most universal significance.

The great secret of the Acropolis cult has been preserved by the Church Fathers in their assault on pagan religion. Despite every claim to her virginity, they tell us, the goddess Athena had in fact secretly been the bride of Hephaistus, and she had had a son by the fire god, the solar serpent-child Erechtheus, ancestor of the Athenians, in whose honour Athenian children were given a golden, serpentine necklace at birth. The fact of Athena's motherhood, however, was only known to the priests and priestesses of the cultus and was otherwise covered over by a story that made Athena not the mother but the foster-mother of the solar child. In this story, the amorous fire god accosts the virgin goddess who successfully repels him but not before the god ejaculates his fiery seed upon her leg. Athena then wipes the semen with a tuft of wool and, throwing it to the Earth, impregnates the Earth-goddess Gaia. It is then Gaia who gives birth to the solar

child—who therefore is autochthonous, born from the soil—
and Athena, victorious in her virginity, Athenas and nurtures
the son as his foster-mother. It is a myth of surrogate motherhood. The Athenians habitually regarded themselves as earthborn and Athena as their foster-mother and patroness, but this
Athenian patriotic mythology was centered on the ambivalence
of surrogacy and on hiding the 'secret motherhood' of supposedly virgin Athena. The Roman Vesta cult was a duplication of
the Acropolis fire cult but while in the Roman derivative we find
a virgin priesthood, in Athens the virgin goddess Athena was
attended by married priestesses, guardians of the virgin's 'secret'.
Various aspects of the rituals of the Acropolis enacted or symbolically alluded to this 'secret'. In essence, all of this is a variation on the primeval Sky-fire-god's nuptial with the Earth
mother, with Athena as an air and cloud goddess—thus the tufts
of wool[1]—distributing the Sky-god's seed. We see here the most
primitive motifs of Attic agricultural mythology. A yet more
basic configuration is simply Sun-Moon-Earth, with the Sun as
the fire god, Gaia the Earth goddess and Athena the dual-natured
moon goddess who rules over the winds and rains.

This mythology is transposed into a philosophical elemental
cosmology in Timaeus of Locri's exposition on the Panathenea.
The mode of transposition is plain. The citizens of Athens, we
are told, are the children of Hephaistus and Gaia. In Timaeus'
contribution to the 'feast of discourse'—it is actually his purpose
to bring the first human beings into creation as part of an
ensemble of speeches with epic scope—this becomes the doctrine that Fire and Earth are the primeval elements.

> Now that which is created is of necessity corporeal, and also visi-
> ble and tangible. And nothing is visible where there is no fire, or
> tangible which has no solidity, and nothing is solid without earth.

---

1. Wool = clouds. This equation is essential for understanding much of the
mythology and cult associations of Athena. Note also that this symbolism
underpins Aristophanes' satire of Socrates and his followers, *The Clouds*, in allusions that usually escape commentators.

Wherefore also the God in the beginning of creation made the body of the universe to consist of fire and earth.

In the Athenian mythology Hephaestus and Gaia are the primeval parents. Timaeus then informs his Athenian listeners that Fire and Earth are the primeval constituents of creation in a direct parallel. But he then explains that there must be a middle term between these two poles, and for structural, mathematical reasons, this third term must itself be dual in nature:

> But two things cannot be rightly put together without a third; there must be some bond of union between them. And the fairest bond is that which makes the most complete fusion of itself and the things which it combines; and proportion is best adapted to effect such a union. For whenever in any three numbers, whether cube or square, there is a mean, which is to the last term what the first term is to it; and again, when the mean is to the first term as the last term is to the mean—then the mean becoming first and last, and the first and last both becoming means, they will all of them of necessity come to be the same, and having become the same with one another will be all one. If the universal frame had been created a surface only and having no depth, a single mean would have sufficed to bind together itself and the other terms; but now, as the world must be solid, and solid bodies are always compacted not by one mean but by two, God placed water and air in the mean between fire and earth, and made them to have the same proportion so far as was possible (as fire is to air so is air to water, and as air is to water so is water to earth); and thus he bound and put together a visible and tangible heaven.

Thus Timaeus gives us the classical and ancient arrangement of the four elements, arranged in order of rarity to density: fire, air, water, earth. To Athenian ears, already alert to the Hephaestus/Gaia = Fire/Earth equation, the further identification of Air and Water as Athena and Poseidon is obvious, if not explicit, and just as obvious that the remarkably masculine character of their goddess, like the dark and light of the Moon, is a manifestation of the same duality of the middle term. Timaeus' exposition of the elements then becomes an illumination of the inner dimensions of what this pious, Athenian mythology is actually about; its true content is a sacred understanding of the origins and nature of the *kosmos*, from its metaphysical roots to its most basic physical components. Moreover, the four element theory characteristic of the Sicilian and Southern Italian philosophers and the doctrines of proportions enunciated by the Pythagorean sages of those regions is revealed as only another way of describing those things which the rites of the Acropolis enact in ritual and myth. In the 'feast of discourse' the Athenians contribute myth and legend and pious gestures: the Locrian brings philosophical illumination to this. This is how the teachings of the *Timaeus* should be understood, namely against a mythological and ritualistic background much of which is only alluded to in the text. But the more we understand of this mythological and ritualistic background the more the cosmology of *Timaeus* is itself illuminated to modern readers.

An example of this is Timaeus' description of the mysterious 'Receptacle of Becoming' which the speaker himself admits is difficult to describe and which readers invariably find opaque and confounding. Timaeus has several attempts at defining this 'Receptacle', employing numerous analogies, none of which seem particularly illuminating. The Receptacle is compared to a mother, firstly, but then to a nurse, and then, on a completely different tack, to the metal gold. Modern commentators will explain that Plato was himself confused about this, since these analogies seem arbitrary and contradictory, as if the author is struggling with an idea that is only half formed. But the source and also the coherence of this conception become evident when the text is considered in its Athenian, mythological context. While the Demiurge is modeled on Hephaestus, the perfectly passive Receptacle is analogous to Gaia, and we suddenly recognize in Timaeus' descriptions of the Receptacle allusions to her cult and mythos. She is the passive, maternal womb that receives the divine seed, who gives birth to the children of the soil and thereafter nurtures them on her bounty. The interplay of mother-nurse is the cultic ambiguity of Gaia-Athena in surrogate motherhood. In a parallel use of symbols, Timaeus at one point describes the human brain as *ploughland* surrounded by a stone fence (the skull) in which the Demiurge sows divine, celestial seeds. The source of the gold analogy applied to the Receptacle also becomes clear: it is to be understood as an emblem of autochthony, for the autochthons are Hesiod's Golden Race, as Plato tells us in the *Republic*, and metallic gold comprises the solar seeds of the Golden Race slowly incubating in the womb of the Earth. The background to these otherwise odd aspects of Timaeus' exposition is again the Athenian autochthony cult.

This is also the key to understanding Timaeus' descriptions of human anatomy and physiology that, in fact, occupy the larger part of the text. One must remember that it is the primordial autochthon that he is describing, the original plant-man and solar child, Erectheus. This is why, for instance, Timaeus gives no account of the reproductive system until the very end, at which point the genitals are added to the primal man as a type of afterthought and as a consequence of generational decay. The original man was not yet touched by animal reproduction but was autochthonous, born directly from the soil (the Receptacle) from golden seeds in the manner of vegetation. In this respect it must be understood that Timaeus is not describing an actual but an archetypal anatomy and, in this, anatomical analogies with such things as a 'winnowing basket'—a cult object in the Attic Mysteries—provide the relevant clues. It is in these sections of the text that we find Timaeus comparing the human form to an upside down tree. Such an analogy only makes sense in the context of the cult of the primal plant-man. Timaeus draws out, in a philosophical mode, a body of cosmological science that describes the inner dimensions of the Athenian/Attic manifestation of Greek religion.

Finally, a further and still more crucial thing to realize about this aspect of the Timaeus monologue—which in literary terms is deeply sonorous and incantatory in places—is that it is being recited on the Athenian New Year and is itself a ritual invocation of a new cosmic order. In traditional reckonings, where the year is a symbol of the All, the New Year is the day of creation, and thus it is the appropriate occasion for a dissertation such as the Locrian's. But in the plan of speeches set forward for us, it is his task to speak first and to bring the *kosmos* into creation 'by his words, so to speak' and from the *kosmos* bring forth the primeval citizens of Socrates' ideal state, namely an idealized, antediluvian Athens, which citizens will populate the speeches later in the programme. Timaeus is not just to describe the creation, his words are to bring the creation into being. This is an age-old, shamanic pattern; in the rites of the New Year the speaker calls or sings the *kosmos* into creation, being Demiurge to his own tale. The New Year is to the solar cycle what the Day of Creation is to the cosmic cycle. Speaking on the New Year, Timaeus is to call the primordial humans into being just as the Demiurge had done on the day of Creation. This, of course, introduces resonant parallelisms that remove the *Timaeus* from the status of just an ordinary text since author and god are assimilated into the same function and creation and text become parallel, as Timaeus himself hints when he describes the atomic level of his *kosmos* by analogy with letters and syllables. It will be found, in fact, that the text itself is isomorphic to the proportions of the human form, as the French scholar Reme Braque discovered, and so is, literally, an embodiment of the very things it describes. In these ways the *Timaeus* behaves very much as a *sacred* text in the fullest sense—having a microcosmic completeness and adequacy—which indeed it is, but to a religion that is now defunct. The Locrian's purpose was nevertheless to revivify universal meanings beyond the particulars of the Athenian predicament and consequently the text's array of symbols is so primordial and fundamental that it has remained an unsurpassed account of traditional cosmological doctrines. To understand those doctrines we merely need to appreciate its proper mythological and ritualistic background.

The Spangled Tortoise

The Peculiar and Unusual Feature in Hermetic Modes of Exegesis

Introduction

IN this brief article I wish to draw attention to a characteristic of traditional exegesis and symbolism that seems to be rarely appreciated. This characteristic is the importance that should properly be attached to some *peculiar or unusual feature* of that which is subject to exegesis. In fact, I wish to establish this as a principle: that very often in traditional exegesis—whether it be of a text or of an image or of some other order of things—the key to proper interpretation is to be found in the *peculiar and unusual feature*. This is a mark of uniqueness. It is the peculiar or unusual detail, perhaps the unaccountable adjective, that exposes the chain of associations necessary for the unfolding of the inner dimensions of that which is being studied or contemplated. The modern mind tends to skip over or explain away such details. In traditional modes of exegesis these seemingly insignificant and incongruous details trigger a transformation of understanding.

Texts

There are countless literary examples that come to mind. In the Homeric Hymn to Hermes, to cite a particularly elegant one, we find a description of Hermes' antics with a tortoise, the shell of which the young god eventually turns into a seven-stringed lyre. Modern classical studies of this ancient Greek text are either unable to make any sense of this at all or they resort to bizarrely trivial explanations such as that offered in the Loeb edition of the Hymn, namely that the tortoise was regarded as some form of 'good-luck' among the Greeks. The eye of the traditional reader, however — or the ear of the traditional listener, since this was first an oral Hymn before being written down — fixes upon a detail that the modern classicists treat as a mere annoyance, namely that the shell of this tortoise is described as 'spangled'. This word is elsewhere used to describe the starry heavens. Here — unaccountably, it seems — it is used to describe the shell of the tortoise that Hermes turns into his seven-stringed lyre. The classicists throw all manner of interpretations at this irritating detail, trying to dislodge it or explain it away, but to the traditional reader it is the key he or she were waiting for. It suddenly becomes blindingly apparent (to use a phrase befitting Homer) that in this Hymn the tortoise is a symbol of the *kosmos*, much as it is in the Chinese tradition, and that when the god turns the shell into a seven-stringed lyre, the strings are transpositions of the seven planets, and the whole Hymn becomes an exposition of the Pythagorean and Hermetic theme, the Music of the Spheres. It is the peculiar and unusual detail in the description of the tortoise that both triggers and confirms this interpretation. And this in turn becomes the key for the clarification of scores more peculiar and unusual details later in the same Hymn.

Other Homeric literature is the same. The *Odyssey*, especially, is full of such seemingly unaccountable peculiar and unusual features. And thus too Greek mythology in general. In fact, any mythology, for it is a characteristic of myths in general, not just the colorful myths of the Greeks. Typically, it is the strange, odd detail that 'gives it away' or, to resort to a more traditional but still current metaphor, it is the 'loose threads' that unravel the warp and weft of the fabric. These loose threads in the weave of traditional stories and myths are vital to anagogical hermeneutics. Needless to say, the modern academic mind will have none of this and accuses traditional exegetes of engaging in some sort of game, importing their structures into the text, latching on to and exaggerating the importance of flimsy details while missing such vital considerations as the 'socio-economic context' and so forth. There is indeed something playful and game-like about this aspect of traditional exegesis—it involves an intellectual delight and playfulness that is conspicuously lacking from the modern academic milieu—not to mention a sense of humour. The peculiar and unusual feature is often funny and is sometimes absurd. One must engage with the text in the right spirit to participate in the game. This is as much as to say, 'Those with ears to hear, let them hear.' The peculiar and unusual feature is a key into an esoteric dimension of the text that is not explicit on the literal level. What the modern critic fails to appreciate in this mode of exegesis is that the chain of associations that opens out from the text and the depth of meaning revealed in the text are so bountiful, that in traditional exegesis even some mild violation and reconstruing of the literal text is permitted if one must sacrifice a lesser meaning for a greater. But textual contortions are often unnecessary. None are needed in our example of the Homeric Hymn to Hermes. It is plain enough, and immediately the full depth and richness of the Hymn, as an esoteric text, becomes obvious. The 'spangled' tortoise shell is the loose thread that unravels the inner meaning of the whole work, and it illuminates the whole work so that it shines from within with profound meanings, rehabilitated from the dusty mausoleums of the 'classics'.

Iconography

We also witness the peculiar and unusual feature in traditional iconography. There are many examples in Christian art. Here we need to distinguish between the clever games of Renaissance art and a more traditional order of Christian iconography. We are not thinking of the clever allusiveness and 'secret meanings' in Piero's Flagellation. Rather we are thinking of the icons of Our Lady of Perpetual Succour, quite widespread, in which the child Jesus in the arms of the Madonna is mysteriously losing a sandal from one foot. This is the peculiar and unusual feature. In all other respects these icons are a straightforward rendering of Madonna and Child. But, unaccountably, Jesus has lost or is losing one sandal. Why? There is a deep and profound symbolism attached to this 'loose thread' that awaits those that care to contemplate it. It is a simple detail but of the same order as the 'spangled' tortoise shell. Such details often seem very clumsy. In the less subtle cases of medieval art it may be simply a case of making one character in a painting much bigger than the others, or giving them a different nimbus or a distinguishing coloring of red and blue over and undergarments. This is a vocabulary of symbols and symbolic devices, and one of the uses to which it is put to is to leave 'clues' to deeper meanings in what seem strange and incongruous details. So-called 'occult' illustrations in later times exulted in this device, but increasingly as an empty gesture.

Astrology

In this article, however, I want to suggest that we meet the peculiar and unusual feature in areas of exegesis beyond text or image, even to the extent that it appears to be an hermetic function, woven into nature as much as in the sacred orders revealed to man. It is not merely a game devised and played by writers and readers of arcane texts and painters and viewers of religious icons, but rather something more integral. And as a principle, and as a tool of exegesis, it should be seen as having wider applications. For example, to continue with the 'Music of the Spheres', we meet the peculiar and unusual feature in the modes of exegesis brought to traditional astrology and horoscopy. In the astrological chart of the heavens, regardless of whether we consult it for noble or ignoble purposes, we encounter an array of planetary and other configurations, some of which are common and some of which are unusual and rare. The astrologer cannot make much of the fact that the Moon is in Aries at any given time because the Moon is in Aries once a month, every month. But if the Moon, Mars, Venus and Mercury are all in Aries—that is peculiar and unusual, and the astrologer therefore grants it a greater significance.

All the factors in a chart of the heavens are weighed up in this manner. This is because the astrologer is searching for the unique quality of a particular moment frozen in time. The astrological chart of the heavens is a graphic and symbolic representation of a unique moment, and it is the essence of that uniqueness that the astrologer seeks to divine in his art. But, in any given case, there are so many factors to be considered, so many possible permutations of the data, the multiplicity of symbols becomes overwhelming. Astrology is prone to this. The key to exegesis, then, is to find the peculiar and unusual feature in the case at hand. The astrologer works by considering all the major factors, then other possibilities, attempting to reach a synthesis. But in any given chart of the heavens there will be one thing that stands out, one thing that particularizes that chart. The experienced astrologer will have seen thousands of charts of the heavens. What then is peculiar and unusual about this one? That is always the question to be answered. That is how the astrologer grasps for the Unique. When that is grasped, all the symbols of the chart are illuminated from within by profound significances and an overwhelming internal coherence. It is the same method by which one reads such a text as the Homeric Hymn to Hermes or understands a motif in Christian iconography; by noting the peculiar and unusual feature. It is a sad fact that modern astrologers are too hell-bent on being 'scientists' or at least 'psychologists' to detect any levels of cosmic humour in their horoscopes and nativities, but there is plenty to be found. The ancients described the planet's courses as a race track, but the Sun and Moon are also Punch and Judy, and Venus and Mars fall in and out of love, are faithful and not, by season. One must see the fabric before one can see the loose threads. In astrology's integral form, there is a type of intimate humour and playfulness—Hermetic in principle—that happens between the astrologer and his charts. In some respects, until this quality develops in an astrologer—a sense of the *kosmos*' humour without which one cannot see the peculiar and unusual—he is only an apprentice.

Dreams

Directly analogous to this is the art of dream interpretation. Properly understood, this is not a formulaic or mechanical matter, but a case of learning the particular 'language' of dreams, its grammar and structure and its typical modes and techniques of communication. Here again we find that the peculiar and unusual feature is the key to a great deal of understanding. Needless to say, this is almost entirely a lost art in the modern West but is still to be found in cultures informed by tradition where the dream is an important event and where the truth and power and transcendent origin of dreams is implicit. The dream interpreter, like the astrologer, and like the exegete of text or image, is confronted with an array of symbols and amongst them must find the key, the peculiar and unusual feature. Obviously dreams do not communicate in plain speech. They communicate in a language of symbols, but a key to understanding them is that the weave of the dream will leave loose threads, and it is what is peculiar, incongruous, odd that is most important. Much psychoanalytic theory and method acknowledges this simple fact too—the therapist latches on to the incongruous detail—but often (or even systematically) not the right one. The trained dream interpreter knows what to look for. Often the peculiar and unusual thing—the key—will only be peculiar and unusual in a sequence of dreams, or is subject to a 'pun' or a play on words. Freud's explorations into these modes of the dream were counter-traditional; the 'Freudian slip' (a loose end by which the therapist dismantles your personality) is inverse to the 'peculiar and unusual feature' which reveals an abundance of higher, not baser, meanings. It is necessary to add that this principle of the peculiar and unusual feature should in no way be confused with a certain sociopathology of modernity that, for example, seeks to understand human beings in general by the study of mass murderers, child molesters, urban cannibals and the like. The peculiar and unusual should not be automatically identified with the morbid and perverse. Contrary to modern assumptions, beauty is as likely to be peculiar and unusual as ugliness.

Homeopathy

Finally, let us note another and very precise and pure application of the peculiar and unusual feature in a mode of traditional medicine, homeopathy. Here, as it was reformulated in its modern practice by Samuel Hahnemann, the physician attains a full picture of the patient's symptoms, always searching for that one peculiar and unusual symptom that will lead the physician to the cure. Homeopathy operates on the hermetic parallel between the microcosm and the macrocosm.

The homeopath searches for parallels between the symptoma-
tology of human pathology and the toxicology of natural substances relative to the healthy human organism. The object of
the search is to determine the *similimum*—the remedy with the
toxicology that is the nearest parallel to the symptoms of the
patient. To such a parallel substance, homeopathic theory maintains, the human organism is supersensitive and thus will
respond to it in miniscule doses, the nearer the parallel the
smaller the dose required, even to a point beyond which there
are no physical molecules of the original substance remaining in
the medicine.

But as in astrology, as in rich texts like the Homeric poems, as in
dreams, one encounters a profusion of data, in this case a
profusion of symptoms and an array of substances known to cause
them in a healthy person. Anyone who has ever encountered the
massive homeopathic compendiums published in its heyday
before the ascendancy of modern industrial medicine (allopathy)
can attest to this profusion. In his Organon of Medicine—still the
bible of purists in the homeopathic fraternity—Hahnemann
formulated his 'new' medical science in strict tenets and
explained that the key to finding the *similimum* is to find, in any
given case, the peculiar and unusual symptom. A patient who has
fever and thirsts has nothing peculiar. There are any number of
substances with fevered thirst in their toxicology. But a patient
who has fever without thirst presents with a more useful
symptom. There are fewer substances that induce a fever without
an accompanying thirst. And fewer yet with fever accompanied
by revulsion of drinking. And so on. Hahnemann was ridiculed
for wanting to distinguish between an itch and a tickle, but he was
searching for the strange and unusual symptom. It follows—it
should be noted—that a homeopath must therefore have an
excellent knowledge of what is to be expected in pathology in
order to be able to see what is peculiar and unusual in any given
case. The homeopath sees the symptoms of disease as a language
by which the organism communicates the nature of its
imbalance. The trained homeopath reads these symptoms—very
much, I contend, like an interpreter of dreams—watchful for the

key. When the homeopath finds the peculiar and unusual symptom it will point to one and only one remedy, and upon further investigation it is revealed to be a match for symptoms the patient had not even reported at first. The homeopath has found the *similimum*, the nexus between micro- and macrocosms, by which he can heal. There are no generic medications in homeopathy. Each case is highly particularized. If a dozen patients present with the flu, they may receive a dozen different remedies because the peculiar and unusual symptom in each case has led the homeopath to twelve different *similima*.

The purpose of citing examples from such different endeavors as interpreting an Homeric text, casting a horoscope and homeopathic diagnostics is to draw attention to the range and extent of the application of this principle. The cases of homeopathy and dream interpretation demonstrate how even in the order of natural phenomenon certain keys occur for those with ears to hear and eyes to see. The reason we find it in sacred texts (and it is scandalous that the Homeric corpus is not routinely considered among sacred texts) is that such texts are effectively parallels to the natural order, or rather to the translucid Nature that is the primordial revelation to which this principle is integral. The principle, in short, is that the peculiar and unusual features reveal the transcendent uniqueness of things, which uniqueness is the key to understanding not only the Unique but also the Universal. It may be the dogs guarding the gates of Alcinous' palace realized as the dog stars of Sirius that alert us the astronomical schema of Homer's Phaiacia, and then of the whole *Odyssey*, or it may be a remarkable angular relationship between planets in a geniture, or it may be a pun in a dream, or it may be the shade of blue of the lips of a patient presenting to a homeopath for some seemingly unrelated ailment: in each case nature or scripture leaves keys or clues—certain peculiar and unusual details—to the inner illumination of the order of things beneath the level of surfaces.

As Above, So Below

On Astrological Correspondences

BY virtue of the Hermetic principle 'As above, so below', all things in the sub-lunary realm *correspond*, as we say, with the celestial order of the heavens. This is the first article of astrology. But this cannot be a conclusive and self-sufficient correspondence, for the simple reason that the celestial order is itself, like the things of the sub-lunary realm, *created* and is not, therefore, either a cause or an end of terrestrial manifestation. Whether the 'above' of the stars acts as a *secondary* cause to events on earth 'below' is another matter, chiefly of concern to practitioners of the lower forms of the astrological arts, but, more importantly, all things whatsoever, whether terrestrial or celestial, have their cause and end beyond themselves in the Uncreated, and the axiom refers to this as much as to any causal connection between Earth and Sky. The celestial order, that is to say, consists of contingencies as much as the sub-lunary order, so that the correspondence between the two orders does not have its beginning in either of them and any notion that the stars 'influence' events on earth can only be part—and the lesser part—of the story. Although the Hermetic axiom frames a cosmology—Hermeticism taking the historical form of a cosmology able to naturalize itself in several traditional orders, pagan, Christian and Moslem—there is an *above* that is beyond both Earth and Sky and to which both Earth *and* Sky are therefore *below*. Beyond the cosmological correspondence between Earth and Sky the axiom refers to a metaphysical truth, namely the correspondence between the Uncreated Principle and all its manifestations, or between the Divine and the mortal, the Uncreate and the created, the Unmanifest and the manifest. The Divine is the *above* and the created—*including* the celestial order—is the *below*. It is this correspondence that generates the others. Simply put, God's creations reveal God, and it is only for that reason that one contingent order, by way of participation with God's relationship to His creations, can *correspond* to another, bearing in mind, of course, that, from the 'higher' point of view, God's creations are not God Himself and that God is in no way changed or effected by His manifestations which are only, as Plato put it, the *best possible likenesses* of Him. This is to say that, while an effect has a necessary relation of correspondence to its Cause, it cannot itself be its Cause. The correspondence between the celestial and the sub-lunary realms, therefore, only comes about because the two realms share a common order, and this is because they share a common Source, a single Creator Who transcends them both.

The symbol of this relation in man's visible universe is the relation between the darkness beyond the stars and all visible things. The darkness beyond is *above* vis-à-vis the entire visible realm *below*. The darkness, in this case, represents the Unmanifest and the visible the manifest. This is a simple way of explaining how it is that modern astrologers have forgotten the metaphysical basis of their science; they only see the stars and have no regard for the eternal ground of darkness against which the stars appear. It is in relation to that eternal darkness that the ancient Greek cosmologists said that Hades—the *under*world— 'extends from the milky way downwards', since what is *under* is obviously relative to what one takes as being *over*. There is, in any case like this, a principle of *ratio* and *proportion* involved, and to lose sight of it is to fall into metaphysical confusion and to start supposing that the stars have, in themselves, some miraculous power over the mundane realm. The same comments must be made concerning man when taken as a microcosm. The

Hermetic doctrine is applicable in that instance too, although the more exact metaphor to be employed in that case is 'As without, so within'. Man encapsulates the universe as microcosm to macrocosm, but this can only be so because there is a common point beyond both realms at which that duality is resolved. One *corresponds* to the other because they have the same Cause, the same Creator, Who is *above* His creation *below*.

Another important point that needs to be clarified in regard to this is the relationship between the whole and its parts. It is precisely because God is unaffected by His creations, and by the act of creating them, and so is therefore eternally unmoved and undiminished, that He reveals Himself in both the totality of His manifestations and in every particular manifestation, in the whole and in the least of its parts. This is why, as Blake put it, there is a world to be seen in a grain of sand and a heaven in a wildflower. The particularities of the *kosmos* are only particulars from the human, which is to say the contingent, point of view. From the *transcendent* viewpoint the distinction between universal and particular disappears just as do such distinctions as subject and object. God is One, and if the universe appears to be *many*, it is not so when considered from the perspective of God's Unity, which is indivisible. God is fully Immanent in His creation and the whole of His Being infuses the least part of it. This is in no way a 'pantheistic' doctrine, because its obverse is equally true, namely that God is utterly remote from His creation and cannot be found in either the *kosmos* as a whole or in any of its parts; rather, it is simply to state that God is Omnipresent, which is an entirely orthodox teaching—the heresy of pantheism, as is the nature of heresies, affording this teaching an exclusive truth without proper regard for the paradoxes that are inherent in the human point of view. To conceive of God as Immanent without, at the same time, conceiving of Him as Transcendent is, in fact, to fail to recognize the human viewpoint for what it is and thus rob God of His total Otherness and Incomparability.

To speak of correspondences is nevertheless not only to speak of likenesses but also, at some level, identification. As God is *above*, so is His creation *below*. Plato expressed this by describing the two in terms of divine paradigm and mortal copy, idea and artifact. But for God in His Unmanifest aspect, of course, there is no 'above' and 'below'; no relativisms whatsoever. The only Reality the *kosmos* can enjoy is His. This is the sense in which the whole created realm is *maya*, illusion, for it is Nothing in relation to God's All; yet, to restate the same paradox as before in another way, it is also the sense in which the whole created realm is 'Every-thing' in relation to God's 'No-thing', since the fact that we cannot see God, or touch God, and that He is not obvious to us, but an invisible, incorporeal Being—the fact over which the atheist stumbles—is a necessary quality of God's Creation too. The consequence of this for the doctrine of correspondences is that, finally, all things correspond to all other things, all is in sympathy, just as all the points on the circumference of a circle are, while distinct from one another from one point of view, finally interchangeable with each other because each is nothing more than a 'projection' of the one point at the center. In theory, then, all particular manifestations are interchangeable with all others, all points of Multiplicity are identical to all other points, and in this way they resolve themselves into Unity. We can only say *once more* into Unity, metaphorically. It is only for convenience's sake—because we must speak as creatures—that we speak of sequential events, cause and effect, when speaking of the Divine and it is only for convenience's sake—a purely human convenience—that we differentiate various modes of God's Being, Manifest and Unmanifest, Immanent and Transcendent.

Given that astrology is a wisdom that addresses a cosmic order of symbolism that is especially adequate as a manifest revelation of the Uncreated—although in theory, following from what has just been said, it is, strictly speaking, no more or less adequate than any other in itself, the adequacy referring only to the human predicament—the astrologer must be constantly aware of these doctrines and their implications. In particular, inasmuch as the astrological art involves discerning correspondences, the very nature of *correspondence* and its metaphysical basis must be kept in mind lest correspondences are only half-discerned, as it were, which is the road by which a sacred science such as astrology readily degenerates into a superstition and a sham. To half-discern correspondences is like noticing that two points on the circumference of a circle are akin as points, without also noticing that this is true of *all* such points and without paying any regard to the point at the center of the circle by virtue of which all circumference points are akin, or, to extend the illustration, it is like noticing that two radii of a circle point in a similar direction without also noticing that this is true of all radii and that the similarity exists only by virtue of the fact that all radii converge in a common center where similarity becomes identity. To use a simple example, it is one thing to note a correspondence between the Sun in the celestial order and a particular manifestation in the sub-lunary order such as, say, a sunflower, but it is another to realize what it is that these two things *ultimately* have in common and, as well, to realize that, theoretically at least, *all* manifestations in the sub-lunary order have a correspondence with the Sun, as with all the other planets. It is never finally enough to ask of what planetary nature is such and such a thing, as if the things of the world could be neatly compartmentalized into seven planetary boxes; *similarity points to identity*; the real *correspondence* between the Sun and a sunflower lies in the Being of God and not in the things, qua *things*; as *things*, rather, they manifest only *difference*, which is the difference inherent in manifestation—Multiplicity—itself.

This fact explains the experience of all astrologers at a particular point in their development; there comes a time when similarities appear, often in lightning-fast sequences of associations, everywhere, and the astrologer is overwhelmed by a profusion of correspondences that are so broadly interconnected and widely interchangeable that nothing intelligent can be said of them at all; all is confusion. The astrologer may note, for example, that while the correspondence between the Sun and a sunflower is immediately obvious, on further consideration the sunflower, as seen in its yellow petals, its upright 'dignity' and so on, partakes of the nature and has the qualities of the planet

Jupiter. While these qualities appear to be distinct, he may then note certain things about the flower that could justify describing it as Mercurial, and then as Saturnine or Martial or Venusian or even Lunar. Referring to the sphere of the fixed stars and the twelve zodiacal signs is no better, for sufficient contemplation will eventually reveal that the sunflower 'belongs' to and has the qualities of them all. This experience is only to be expected. After observing astrological correspondences for some time, the astrologer suddenly comes under the sway, as it were, of a centripetal attraction towards Unity and Identity. Suddenly, all he sees are similarities that together lead into an interconnectedness that is very quickly beyond being encompassed with the ordinary operations of the mind. This is not an unusual or a profound experience in most cases; rather, it presents itself as a maze in which the mind may become well and truly lost, a period of confusion in which, since everything corresponds with everything else, there is no point in observing correspondences at all. Since the things of Multiplicity are indefinite in number—the cosmic indefinitude being an image (to continue with Platonic terminology) of the Infinity of God—this is the experience out of which arise various syncretic 'occultisms' that are characterized by a hopelessly confused interconnectedness. On the other hand, it is out of this experience, which is at once a dissolution and a crystallization of possibilities—the *solve et coagula* of the alchemists—that there arises the opportunity for a true synthesis, which is none other than the experience of the One. It becomes all-important, in fact, to pursue all correspondences to Unity, because it is only by knowledge of the One that the reality of a correspondence can be judged.

This brings us to the Unitive experience that is, properly considered, the ultimate aim of the contemplation of the symbolism of astrology and the experience that confers upon astrology its sacred status. In a world in which all things finally correspond to each other, without reference to *that through which they correspond* it is impossible to establish any criterion of truth; on its own and to the everyday consciousness of man the world is an abyss of relativisms that have a semblance of an order that is ungraspable.

J.B.S. Haldane was right when he quipped that 'the universe is not only queerer than we think, but queerer than we *can* think' provided we understand by his words that it is merely *human* thought for which this is the case. The Unitive experience—when the *sameness* that gives correspondences their verisimilitude is realized not in particular cases but in *itself*, when the world is seen in a grain of sand and heaven in a wildflower—necessarily employs a supra-human 'faculty', an *intuition*. The ratiocinative mind cannot grasp the One since the created cannot encompass its Almighty Creator. Only God (Immanent) can know God (Transcendent). Intuition, on the other hand, is defined as the uncreated faculty that perceives the Uncreate. It is the intuition that perceives the Immanent 'now' and 'here' in (and simultaneously 'beyond') the realm of time and space. The more fully developed this intuition, the more able is the astrologer to discern true correspondences. In the case of the sunflower, to continue with that example, it is the ratiocinative mind that is finally led to the conclusion that the flower's qualities are indicative of all astrological symbols interchangeably. It is the intuition—'intuition' because it involves a sudden leap out of the confines of time and space—that is capable of synthesizing this profusion of interconnectedness into Unity, revealing the true uniqueness of the flower, which is its *image* of the 'Uniqueness of the Unique'. Being uncreated itself, however, it should be stressed that it may be misleading to describe this intuition as a 'faculty'; it does not involve in any way any exertion of the human will; rather it is entirely *receptive* and in order to 'develop' it one must submit oneself to the Divine Will and place oneself as a mirror, so to speak, under the rays of Divine Illumination. It is, that is to say, an *inspiration*, and it is finally inspiration upon which astrology depends, even in the art's baser prognostic forms.

The Hermetic axiom quoted at the outset, it will be noticed, explains itself in terms of a vertical symbolism. This, in view of the above comments, deserves some final consideration. If we were to extend it to a symbolism with two coordinates instead of only one, that is, the symbolism of the cross, every 'below' is then a horizontal set at right angles to the vertical axis and in which the vertical axis terminates. If the 'above' is the Creator and the 'below' the spatio-temporal *kosmos*, the horizontal axis of the cross symbolizes the indefinite extension of the *kosmos*. Considered without any reference to the vertical axis that actually defines it, this 'horizontal world' is an endless labyrinth. Man's spiritual realization depends not upon fumbling around in this labyrinth, but in transcending it, which entails movement upon the vertical axis *up*. In terms of the horizontal, the point at which the two axes intersect is a *center*, and developing the intuitive faculty is, considered in terms of this type of representation, a return to that center where, *simultaneously*, the Holy Spirit descends and the soul of man soars heavenward. It is at that central point, which is, of course, immanent to every point in space and every moment in time, that the full implications of 'As above, so below' will be revealed. This type of axial representation is, furthermore, part of the formal symbolism of astrology since these two axes are what the horizon and the meridian represent respectively in the geocentric universe. An astrologer 'reads' a chart of the heavens in these axial terms.

The final truth revealed in any chart of the heavens, such charts encapsulating as they do a *here* and a *now*, is not to be discovered, however, in either axis but rather in the point at which they meet. Just as the darkness beyond the stars is finally the most profound of the celestial symbols, so the most profound of astrology's formal symbols is the point at the center of a chart of the sky. There, in natal astrology, is the true mystery of Personhood to be contemplated. It is the Uniqueness of the Unique and Uncreated point at the center of the chart, and not the other configurations of symbols *that merely point to it*, that the true astrologer 'divines'. Standing contemplating the open heavens rather than a chart, this point is within the astrologer himself and to 'divine' it in that case is to 'Know thyself!' which, as the ancient mysteries taught, is finally to say to God, acknowledging the ultimate paradox of the human state, 'Thou art!'

The Circle & the Square

Signs, Symbols & Glyphs in Traditional & Modern Astrology

Planetary Symbols and Glyphs

THE cosmological, and also the metaphysical, underpinnings of any body of traditional astrological science are revealed in the appropriateness and adequacy of its symbols, including its literal glyphs, signs and sigils. Conversely, corruptions of an integral astrology, which amount to losing sight of the sciences' cosmological and metaphysical foundations, will be revealed in the proliferation of distorted symbols and pseudo-symbols and sentimental and meaningless glyphs, signs and sigils. A spectacular instance of this is to be found in the contrast between the traditional planetary glyphs employed in Western astrology and those attributed to the newly discovered extra-Saturnian planets, Uranus, Neptune and Pluto. The glyphs given to the ancient seven planets have a simple but profound basis of construction. The glyphs now accepted and given to the extra-Saturnian planets, in contrast, are sentimental constructions of only shallow, if any, significance. The ancient seven planets all have glyphs constructed of various arrangements of three basic linear elements—

circle, half-circle, cross—each of precise alchemical significance. They are, amongst other things, the tripartite distinction—*pneuma, psyche, soma*—used by St Paul. The Sun is represented by the circle only modified by defining its center with a point. The Moon is represented by the half-circle often modified to a crescent. Venus is represented by the circle surmounting the cross without modification. Mars is represented by the cross surmounting the circle, usually with the cross modified to an arrow to avoid confusion with the glyph of Venus which is its mirror opposite. Jupiter is represented by the half-moon surmounting the cross and Saturn by the cross surmounting the half-moon. Mercury is a synthesis: the cross is surmounted by the circle, which is surmounted by the half-circle modified to a crescent on its back. The glyphs of the ancient seven planets, that is to say, are a group of permutations of three basic signs, and each arrangement is an alchemically precise description of the character and nature of that planet. The glyphs of Venus and Mars describe the female and male respectively, for example. These are, in any case, intelligent and meaningful symbols, although this is rarely appreciated among modern astrologers. The planetary glyphs did not 'evolve' from 'archetypes' in the 'collective unconscious', nor are they 'abstractions' from naturalistic representations or anything of that sort: they are quite simply a language, an algebra, used to describe cosmological and metaphysical realities, or short-hand marks for a spiritual vocabulary that describes cosmological and metaphysical realities.

This is not true of the glyphs associated with the extra-Saturnian planets. In some cases their attributes are just silly. The glyph for Uranus, for instance, is historically—regardless of what other spurious origins New Agers wish to attach to it—merely a stylization of the letter H surmounting a circle, the H standing for Herschel the astronomer credited with 'discovering' the planet in 1781. Subsequently, imaginative astrologers in the twentieth century, noticing the uncannily prophetic resemblance of this stylized H to the ubiquitous television aerial surmised—therefore!—that Uranus must 'rule' modern telecommunications, TV, radio, sonar, radar, and such scientific inventions in general. In the case of Neptune it was simply a matter of attributing the trident of the sea god to this blue-green colored planet in an unimaginative chain of associations. But subsequently later astrologers—this time during the heyday of Theosophy and seances—leapt upon the similarity of the trident to the Greek letter Psi and from there attributed psychism—it begins with Psi!—to this new planet. Thus is Neptune, in modern astrological orthodoxy, the planet of psychic powers, dreams, hypnosis, the trance state, and so on. It is ignorant and also arrogant to suppose that the ancient mind developed the glyph of Mars from nothing more than the chain of associations starting with Mars looks red, blood is red, war spills blood, Mars is the god of war... The glyphs of the ancient planets are a coherent system and are adequate on many levels, in fact, on all necessary levels. Yet the associations given to Neptune and Uranus in modern astrological practice grow out of nothing more than sentimental and often fanciful chains of pseudo-correspondences that only mimic the meaningfulness of the web of real correspondences woven around the ancient planets since before history. The glyphs of the ancient seven planets are demonstrably intelligent and reveal an intelligent and integral doctrine. The glyphs of the modern planets—the glyph of Pluto is a confused issue but the usual configurations seem nothing more than unintelligent parodies of the glyph of Mercury—are superstitions by contrast.

Square and Circle: Earth and Sky

An even more fundamental contrast is that between the traditional manner of depicting a geniture or horoscope or chart of the heavens and the modern manner which is now used unanimously in Western astrological practice. It is the contrast between a square and a circle. From ancient times, and throughout the Middle Ages, and in fact up until quite recently, the traditional way of depicting a horoscope was in a rectilinear chart divided by a particular arrangement of diagonals.

Horoſcopium geſtellet burcb
Ioannem Kepplerum
1 6 o 8.

The astrologers who reformulated and to some extent reinvented a distinctly modern astrology in the later half of the nineteenth and early part of the twentieth centuries, astrologers such as Alan Leo (1860–1917), championed, amongst many other innovations, the use of the circular chart and discarded the square chart as old fashioned.[1] By the latter half of the twentieth century the traditional square chart had completely disappeared from common knowledge, and today all but a few of those calling themselves astrologers can give no account of how the square chart operates or the basis of its symbolism. The round chart reigns supreme. But the round chart, when employed for a nativity or for some matter of mundane astrology, or for whatever earthly or human purpose, is profoundly—even exactly—wrong in its symbolism. Circular forms, in the most basic vocabulary of astrological symbols, denote the heavens. The Sun is circular, as is the Moon. And they move in arcs and cycles. In the northern skies the constellations circle the pole star. All in the heavens is the geometry of circles. Rectilinear forms—the square—in contrast, denote the terrestrial, the sub-lunary realm. The contrast is so commonplace in traditional thought and traditional art and iconography it hardly needs reiteration; it is as basic as a building, a temple, a mosque, a basilica, consisting essentially of a dome on a box, the sky over the earth.

Now a horoscope that records any event, such as a birth, records, we must recall at this point, two factors, time and place.

1. The modern astrological revival was largely Theosophically instigated and many names could be mentioned. Alan Leo—also Charles Carter and then Margaret Hone—played a major hand in constructing and proliferating modern Western astrological conventions.

A horoscope is a symbolic representation of certain calculations pertaining to an event at a particular time and at a particular place. In theory the positions of the orbs of the heavens—as seen at a particular time at a particular place—encapsulate the unique quality of that event. In theory, in fact, we encounter the paradox that by thus mapping and particularizing an event in time and space, capturing its uniqueness, astrology supposes it achieves insights into the Transcendent, eternal qualities, beyond time and space. The important thing to note, nevertheless, is that *spatial* as well as temporal particularization is absolutely essential. A horoscope calculated for no place in particular is inconceivable. Events happen in space. Every birth happens somewhere, and you cannot cast a horoscope for an event such as a birth without knowing where it happened, nor when it happened—but more importantly for our current purposes, *where*. This fact is represented in the traditional chart by recording that the event happened *in the terrestrial realm*; thus the chart is square. The round chart, on the other hand, implies, by this symbolism, that the event occurred in the heavens. There is an obvious confusion here. The event that occurs in the heavens is a certain arrangement of the stars. The round chart depicts that. But seen from where? The square representation follows a different accentuation of the same symbolism, or actually it depicts a different event. The square chart does not depict the event that occurred in the heavens but an event that occurred on Earth, namely that *someone consulted the heavens or an event occurred* at a particular time and place. The square chart does not record the positions of the orbs of the heavens, but rather their positions *as seen from the earth*, and in particular as seen in the context of the four cardinal directions that define the sublunary realm. This is not to say that the round chart is not geocentric, but it records the arrangements of the stars from an *abstracted geocentric* view point, so to speak.[2]

---

2. The abstraction is very like the abstraction we see in photographs of the Earth from outer space. Such pictures help us forget and abstract us from the rectilinear reality of human terrestrial existence as defined by the cycles of day

The Mystery of the Center

We can see this by considering the symbolism of the *center* of the horoscope. In the round chart method the central point is exposed and in graphic terms anonymous since the name and other information is usually left away from the round chart to facilitate seeing 'aspect patterns', another feature of modern astrology emphasized by Alan Leo and the other 'pioneers' of modern astrology. In the square chart we find the central point covered by a square inside the square, and in this space is written the details of the event that is the subject of the chart, the event that precipitated the inquiry. The central point is obscured by the record of mundane events, time and place. Here we encounter the graphic correlative of the paradox mentioned above, and the square chart expresses it eloquently and exactly while the round chart is both inappropriate and inadequate. The central point is obscured in the square representation because it represents the Transcendent. Again: astrology supposes that by particularizing an event in time and space, capturing its uniqueness, its 'Now', it achieves insights into the Transcendent, eternal qualities, beyond time and space. By knowing the unique you will know the Unique. But no one should suppose this is easy. It is a paradox. A problem. Because by grasping at a moment in time and space you seek what is eternal and beyond time and space. In the square chart the profound uniqueness of the recorded event—the center—is hidden by its mundane details. In a nativity the true secret, the 'meaning' or

and year. This rectilinear framework is the universal experience of peoples sufficiently close to the reality of the Earth and of nature: only modern, Western man has no sense of it at all and treats countless centuries of human wisdom with contempt in a glib phrase about how science has proven the world is round, not flat as the Church maintained, and so on. That the world is round when viewed from the 'outside' only underlines the fact that it is rectilinear when viewed from the 'inside'. But in a very real sense modern man has lost the 'inside' view and goes about 'inside' the world with a view from the 'outside'. This is exactly what happens when we draw a round astrological chart for a terrestrial event.

ultimate, eternal significance of an incarnation is obscured by our name, date of birth, place of birth, the identity we are given, the circumstances into which we were born. The problem in natal astrology is divining what is *behind* the central box. In personal terms for the native who is subject of the inquiry the problem is how to discover that 'true self'—the center—hidden behind their name. In reincarnationist applications we find that all incarnations share the same center, the same Transmigrant. The center is Unity, One. The astrologer who casts natal horoscopes over a period of time should realize that the center of every chart is the same. It is the only thing that is the same in every countless unique configuration. There is only one center.

In the round chart it is as if our true being is on display, as if there is no paradox. In short, the round chart gives us a false view of the human predicament. Its symbolism, in fact, would be more appropriate for *gods*, of celestial birth (in the realm of circular forms) and wearing no mask, no veil (the Transcendent center point uncovered). The symbolism of the square chart implies such traditional themes as mortal life (conceived as one or many) being a prison—the square is a box. In this metaphor the key is in the center of the chart but hidden from you, and the paradox is that it is yourself or the person you understand yourself to be, your mundane identity, that is hiding it from you. All the various configurations of the planets in signs in a horoscope point to the (hidden) center, and derive any significance they may have from this and this alone. As the astrologer synthesizes these signs and configurations and their interconnectedness he is really divining the uniqueness of the Unique, the veiled center. The error of the round chart mode is this: the center cannot be—or cannot only be—representative of the event/observer point in time and space. For where then is the Transcendent point beyond time and space? There must be both. There must be that which signifies the terrestrial event and there must be a point signifying that which is beyond time and space. Yes, the center *is* both, but *not from the human point of view*, only from the privileged viewpoint of gods. Put more severely, the round representation implies there is no *maya*, no illusion, no schism of the center for man. If we remove the inner square from the traditional square chart to reveal the center then we are in the foursquare Garden of Eden at the 'tree in the midst' and the prison turns into paradise. Put more severely still, the round chart of modern astrology is a symbolic representation of the *hubris* that man is not a fallen being alienated from his true Self; it is a symbolic embodiment of the modern sentimental self-flattery that regards ego-bound man as a self-sufficient being. The square arrangement, in contrast, embodies a precise and coherent symbolism adequate to the human situation and appropriate to the purposes of astrology. The difference between representing a chart of the heavens as a square or as a circle is not just a matter of taste and aesthetics;[3] there are fundamental differences of symbolism.

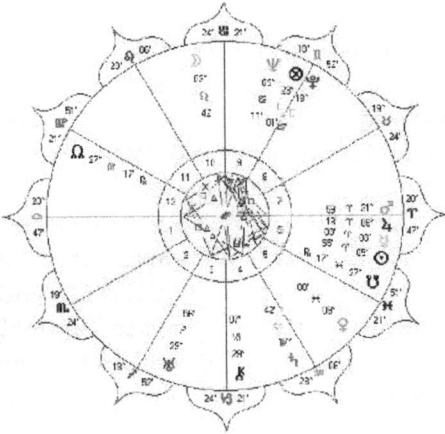

A typical variation on the modern, round horoscope (in this case a nativity attributed to Joseph Campbell). The reader will note the mess of 'aspects' recorded in the center of the chart. The pseudo-lotus petal form given to this chart is very common but utterly meaningless; it only serves to give the impression of the 'mystic east'.

3. The popularity of the round chart is to some extent aesthetic. New Age sentimentality prefers circles to squares.

The Horizontal and Vertical Axes

The other fundamental problem with the round chart, other than overlooking the rectilinear nature of the sub-lunary realm and failing to express the mystery of the center, is that it is based upon an essentially horizontal metaphysical symbolism and in theory and in practice obliterates the significance of the vertical axis. In astrological terms, it emphasizes the Ascendant/Descendant axis at the expense of the Midheaven/Immum Coeli axis. It is simply a case of the square chart being obviously founded upon two axes, horizontal and vertical, and therefore representing and bringing into play all the symbolism of the cross,[4] in this case the celestial cross inscribed by the extremities of the Sun's daily and annual course. The same cross can be seen in and expressed in a circle, of course, but the very nature of the circle does not speak of fourness as does a square. In practical terms the round chart lends itself to an emphasis on the horizontal axis. It is no accident that the introduction of the round chart was accompanied by the introduction or revival of the so-called Equal House method of 'Mundane Houses'. In this method the astrologer begins at the horoscopic point—the 'rising sign'—and counts off twelve 'houses' of thirty degrees each. The actual midheaven—the highest arc of the Sun at that place at that time, projected onto the ecliptic, representing 'up'—may or may not correspond to the cusp of the Tenth House. In most nativities and charts, in fact, it will not, and so the vertical axis, the point signifying 'up' in a chart of the heavens, is adrift from the structure of the Mundane Houses.

The whole question of the Mundane Houses is profoundly confused in modern astrology; the 'Equal House' system of defining these 'Houses' (or Temples, in some ancient parlance) seems a simple solution to a complex issue. The issue is less complex when using the square chart. The square chart implies in its structure—in a way that the circular chart does not—the essen-

---

4. This is what is really at issue. The square arrangement makes the symbolism of the cross obvious; the round arrangement obscures it.

tial fourness of astrological coordinates, the fourness created by the horizon and the meridian. The square chart, that is, lends itself to so-called Quadrant methods of House division, where the Mundane Houses are (usually) of unequal size, calculated in quadrants defined by the two axes, horizon and meridian, horizontal and vertical. In this case the cusp of the Tenth House *always* corresponds with the Midheaven and the structure of the Mundane Houses is defined as much by a vertical axis as a horizontal.[5] The question of Mundane Houses is important because it follows from points made above. The square chart, we said, does not, strictly speaking, represent the position of the stars, but rather their positions *as seen from a particular place at a particular time*. The square chart, in its whole character, acknowledges that the event that occurred—someone consulted the oracle of the heavens—was a sub-lunary event. In modern astrology no one is quite sure what the Mundane Houses are, what they signify and why they are important. But it is implicit in the traditional quadrant-style square chart. The point is this: in astrological terms we do not know the heavens in their circular perfection but rather through the rectilinear focus of the sub-lunary realm which is defined by the four cardinal points. The signs of the zodiac do not 'influence' us directly but rather are *mediated through the rectilinear sub-lunary realm*. The circular geometry of the heavens becomes the rectilinear geometry of the mundane world. The circle becomes a square. That is precisely the human predicament and the human problem expressed popularly as the 'squaring the circle' conundrum. Astrology expresses this by creating a 'square zodiac', a 'zodiac' defined by the two axes and their cross of cardinal points, that is a mundane or *sublunary duplicate of the celestial zodiac*.[6] This duplicate twelve-fold division of the space of

---

5. In the ancient world the exact method of calculation, preserving the horizon/meridian axes, has been recorded for us by Porphyry. It is not widely known among modern astrologers even though it is by far the most satisfactory solution to modern debates about methods of House division.

6. In ancient sources the Houses are not always a duplicate of the celestial zodiac but are rather described as various houses or rooms or temples in the Underworld divided in many different ways. Strictly speaking, it is wrong to

the sub-lunary realm, the Twelve Mundane Houses, is highly particularized to a time and place. The houses expand and shrink according to latitude, time of day and time of year. Their cusps move by a degree of arc every four or so minutes. It is to construct this duplicate dodecary of 'Houses'—distorted, let it be noted again, by the 'pull' of the four cardinal points, the rectilinear focus of the mundane predicament—that it is necessary to know the time of an event as nearly as possible. Even a small error can cause large differences. In any case, it is *through* these twelve mundane 'Houses', calculated as precisely as possible, that the pristine (circular) celestial zodiac 'influences' us. In natural terms, there are not just Earth and Stars. There is also atmosphere, the intermediate realm of Air, which, though part of the sublunary realm, shields the Earth from the fire of the Heavens and so is like a third factor.[7] Modern astrologers do not understand why ancient astrological texts dwelt upon weather forecasting. They think it a trivialization of the art. But ancient astrologers were as much *cloud watchers* as star gazers, and there is

---

think of the Houses as another zodiac—that is the thinking that leads to the Equal House solution whereby the equal signs of the Zodiac are simply transposed onto the Houses. Unlike the twelvefold zodiac, House 'systems' can quite legitimately consist of only four houses, or eight, or twenty-four, or other divisions. These are common in ancient sources, the twelve-fold division appearing comparatively late. But whatever division is made, the Houses remain in essence a sub-lunary reorganization of the zodiacal archetypes. In many ancient sources the sub-lunary realm is being conceived as the 'Underworld'. Such sources usually emphasize the importance of the Second and Eighth Houses as gateways or doors between one world and another. It is one of the keys of 'interchangeable symbolisms' that the sub-lunary realm is actually the Underworld (Hades) from a certain point of view. The thinking behind the House divisions can be found in texts like the *Dream of Scipio* better than in explicitly astrological texts like Ptolemy.

7. In mythology we often find intermediaries between the Sky God and Earth Mother. In classical Athens we find Athena as foster-mother between Hephaestus and Gaia, for example. This is an important example because this mythological configuration underpins Plato's *Timaeus*, a seminal text in the Western astrological tradition. Note in that dialogue Timaeus' arguments for why Air and Water must mediate between Fire and Earth and the further implications of that to his cosmology.

an important symbolism in this that is now totally lost in modern astrology. Ancient astrologers understood that the stars do not 'influence' us directly, but through the 'Air' or 'Aether', or rather *the space of the sublunary realm* that is qualitatively sensitive to such celestial influences.[8] The square representation of the horoscope or nativity or chart of the heavens is exactly a graphic symbol of this space. It is then divided into four quadrants at North, South, East and West, and these are then divided into three parts making twelve parts in all. It is as if this is how the celestial zodiac (of the circular realm) is imprinted upon the space of the sublunary (rectilinear realm) at that time, at that place.

In terms of 'reading' or divining or 'analyzing' a horoscope, we understand that the signs of the zodiac 'work through' or 'manifest' in the Mundane Houses. Thus if Sol is in Leo, making for a regal nature, it makes all the difference if, at that time, at that place, Sol was in the first Mundane House, the House governing physical appearance: the native may well have the demeanor of a king but no other regal attributes at all. In another configuration, only hours later, Sol will be in an altogether different House. The Zodiac without the Mundane Houses is remote and abstract, the realm of the gods. The Mundane Houses are the particularization of the zodiac in time and space—*through the focus of the four cardinal directions*—in the sublunary realm. A proper understanding of this makes the adequacy of the square chart plain. By contrast, the Mundane Houses, and the Midheaven's vertical axis, fit uncomfortably into the abstraction of the round method. The round method lends itself to the displacement of the Midheaven, the fracturing of the integrity of

---

8. Which influences are symbolized as 'clouds'. Clouds, that is, gather stellar influences and then distribute them over the Earth. It is important to realize that in traditional—indeed, primeval—mythology it is not that the Sky fertilizes the Earth; more typically, the Sky-God deposits his seed in the clouds, and from there it fertilizes the Earth. The medium of stellar or Uranic influence upon the Earth is a vital issue. This medium is in fact qualitative space. In traditional astrological practice this takes a meteorological symbolism which begins with the four directions correlated to the seasons.

the Mundane Houses, and the mechanical extension of the horizontal axis through the space of the sublunary realm by the so-called 'Equal House' system. The Equal House system, and the round chart, are well-adapted to describing the 'psychology' of the denizens of the horizontal world of modernity but together they violate the ancient doctrines and foundations of astrology. Quadrant House systems that preserve the integrity of the vertical axis, and the square chart, together express an integral traditional symbolism that is metaphysically exact and adequate. In modern astrology, the Midheaven is said to represent such things as the native's 'career'. In traditional astrology it represents rather the inherently spiritual idea of *vocation*. This illustrates the differences perfectly. The modern idea of 'career' is the horizontal counterpart of the vertical idea of vocation. The modes of modern astrology are adapted to a horizontal world in which genuine spiritual yearning is not the norm. The traditional modes are adapted to a world where men were more concerned about the welfare of their soul than the size of their life insurance. In traditional astrology—even in the decadent late classical syncretism of Ptolemy—the Midheaven (and the up-down axis) is more important than the Ascendant (and the left-right axis). In modern astrology the Ascendant, the 'Rising Sign', is of paramount importance. The change from the square to the round chart accommodates this change of emphasis; the round chart is the perfect graphic vehicle for the change.

Space

It has been important to stress the spatial character of the Mundane Houses because, since the Enlightenment, this has been, as already stated, a matter of confusion among astrologers in the West. In fact, since the Cartesian revolution, at least, the West has increasingly regarded space as merely quantitative 'extension' and not as qualitative. Since the Enlightenment, though, this error took hold in astrology such that the deviation of Western astrology from traditional principles has largely taken the form of 'forgetting' the symbolic nature of space. In the early eighteenth century a monk named Placidus, departing from all previous understandings, and acting on an inspiration that is difficult to fathom,[9] developed the notion that the cusps of the Mundane Houses should be determined by dividing the *time* it takes for the sun to traverse through the quadrants. This innovation was vigorously opposed by astrologers of the day but Placidus' system gained popularity through the wide dissemination of easy-to-use tables plotting the Placidean cusps. In fact, it is nothing less than extraordinary—almost like the product of a sinister design[10]—that for nearly a century, the Placidean tables—based on temporal divisions rather than spatial ones— were the *only* tables available to European and American astrologers and so, for several generations, Western astrologers came to consider the temporal method of division the norm and the traditional, spatial understandings were lost. In many if not most cases astrologers labored with 'Tables of Houses' without even realizing they were the outcome of Placidus' innovation or without appreciating alternative and more symbolically correct systems. This must be seen as part of the modern West's general deviation from traditional understandings and specifically as part of the West's loss of traditional understandings of the qualities of space.[11] The Placidean error, perhaps more than any

9. Presumably under Cartesian influence, it seems that for Placidus a purely qualitative conception of space led him to devise an entirely new and unprecedented method of calculation. In any case, the traditional spatial divisions must have already been problematical to Placidus.

10. Why and how and by whose instrumentation the Placidean innovation came to monopolize Western astrological practice and publishing for so long is a matter that deserves a thorough historical investigation.

11. It should be added, as an historical note here, that some aspects of the modern Western astrology, such as the use of the round chart and the Equal House system, found their nurture among the astrologers of Islam and amidst the quite abstract, mathematically based astrology developed by Islamic civilization in the Middle Ages. This is another case where Western 'occultists' have imported ideas, out of context, from their cultural shadows, the 'infidels', and where Islamic modes of abstraction have upset the more concrete forms of the West when the West acquired them recklessly. Islamic astrology demonstrates

other factor, helped to 'horizontalize' the Western astrological mind.[12] The obliteration of a qualitative notion of space was, of course, a prerequisite for the admission of the extra-Saturnian planets into the astrological *kosmos*, as if the celestial sphere envisaged by the purview of the observing geocentric (and microcosmic) consciousness were of no cosmological significance and indistinct from the 'solar system' at large![13]

In practice, the square chart with its system of diagonals emphasizes what are called the 'planets on the angles'; astronomically speaking, the planets either rising, setting or culminating. In the square arrangement you can see at a glance which heavenly bodies were rising, setting or culminating at that time, at that place. In traditional astrology these positions are seen as of

---

to us that such 'horizontal' methods can find a place in an integral cosmology— the horizontal axis has a more obviously profound meaning in Islam as anyone who has sat on the floor in the emptiness of a large mosque will readily appreciate. It is the modern West that succumbed to the 'gravity' of the horizontal axis and it is not surprising to find that an organization of symbolism in modern astrology gives it expression. Nor is it surprising that it is a case where things that were benign in the Islamic order created upheavals to traditional understandings in the West. When considering the 'occult' or cosmological arts and sciences, such as astrology and alchemy, we cannot ignore the fact that the Christian order, intimidated by the intellectual superiority of the Muslims' in the Middle Ages, set about trying to absorb and appropriate 'Arab learning'. In many instances such transplants from one order to another resulted in profound disturbances to the spiritual equilibrium of the West.

12. It certainly prepared the way for the reintroduction of the Equal House system which purported to simplify a complexity only created by Placidius in the first place. Most contemporary astrologers recognize Placidius as wrong, but the damage has been done, and few can be bothered discovering what he was so wrong about and what other raft of errors accompanied this particular error or of what greater malady it is symptomatic.

13. The issue of the extra-Saturnians cannot be separated from other broader issues raised in this article. In modern astrology the extra-Saturnians are like seductive mirrors into which one can project all one's self-delusions. They play a very important function in dismantling traditional understandings in this respect. It is the author's personal observation that of the four of five genuinely gifted and sensitive astrologers he has met, people with a genuine capacity to intuit the symbolism of astrology, and with an inherent depth in their understanding, have all been hopelessly and utterly obsessed with the

utmost importance; it is the *places of transition* that are crucial. You can see risings, settings and culminations in the round arrangement too, but this is not emphasized. Instead, as noted above, the round chart allows an emphasis on 'aspects' by which modern astrologers mean geometrical relationships between the planets in their celestial longitudes. It is always a matter of surprise and incomprehension when modern astrologers discover that astrologers in the past paid little heed to 'aspects', designating them as 'bodily' relationships of no great importance. Instead, the 'aspects' referred to by the ancients are simple relationships between Houses (or Temples). If Sol is situated in the First House and Luna in the Third, they have an aspect of 60 degrees or 'sextile' which is counted as felicitous. If, however, Sol was in the First House and Luna in the Fourth, the angular relation between the Houses (and therefore the planets within those Houses) is 90 degrees or 'square' which is counted as unfortunate. This system, it will be seen, is another application of the 'round' versus 'square' symbolism. The sextile aspect is felicitous because the 60-degree angle resolves itself within the circle in the form of a hexagon which is seen as a 'circular polygon'. In contrast, the 'square' aspect is unfortunate because the 90-degree angle resolves itself within the circle as a square, a rectilinear polygon.[14] The sextile affords an escape from the rectilinear. The

extra-Saturnian planets. The function of these planets is precisely as distraction. Their purpose is to seduce into fantasy any intelligent mind that might stray into modern astrology. The distinction between the traditional seven planets as 'individual' and the extra-Saturnians as 'collective' has no basis in tradition and is merely a reflection of the colorings of 19thC political liberalism imposed upon astrological frameworks on the basis that the discovery of the extra-Saturnians coincided with the rise of socialism. It is also worth noting that the cult of the extra-Saturnians has proliferated in recent times with countless new 'planets' and 'planetoids' cluttering up modern horoscopes as astrologers impart significance to every item of space debris they can fix their telescopes upon. The *'kosmos'* should not be confused with the 'solar system'. Cosmology should not be confused with cosmography. The present author has been surprised to find the extra-Saturnians given uncritical legitimacy in the astrological writings of Whitall Perry.

14. Note the application of this geometrical symbolism in the association of

square aspect is unfortunate because it provides no means by which the native (if the chart being considered is a nativity) can escape the 'box' of mortal existence, no means by which the square can be transformed into the circle. This system is properly attached to the Houses, not to the 'bodily' positions of the planets. Again, the same distinction as between the square and round charts prevails. The modern astrologers who popularized the round chart were consistent in their logic and brought to the forefront the 'bodily' aspects of the planets, which is to say the planets considered in their 'raw' longitudinal positions in the celestial and circular zodiac, unmediated by the Houses. But even in that they furthermore overlooked all that the ancient sources had to say about the importance of *latitudinal* positions in 'bodily' aspects, giving sole importance to longitude—another case of marginalizing or obliterating 'vertical' symbolism, because, by transposition, latitude is the vertical co-ordinate and longitude the horizontal. Strictly speaking, in traditional astrology, it is the Houses that have aspects to one another, not the planets, or rather the 'bodily' aspects of the planets are clearly seen to be factors of a different and in some respects a lesser order. The relations between Houses are good or ill depending on how 'locked in' to the rectilinear 'prison' of the fourfold sublunary realm the relation is. All that is counted as felicitous in traditional astrology concerns regeneration from a rectilinear predicament to the formal perfection of the circular realm; the round chart and 'bodily' aspects by definition dislodge and do violence to this whole symbolism.

Conclusion

None of this is to say that astrology was a pristine, healthy, robust traditional cosmological science prior to modern times,

---

the octagon with baptism and regeneration. The octagon is really two squares superimposed depicting the rectilinear returning again (regeneration) to circular forms.

but until the modern revisions it could be said that astrology's most central symbols—its algebra of symbols—was largely understood and intact, regardless of how debased and decadent the application of those symbols. It can happen in traditional orders that metaphysically or cosmologically exact symbolisms perpetuate from generation to generation despite and not because of the purposes to which they are put,[15] and until recent times even the shonkiest of astrologers was, unwittingly or not, dealing with an integral symbolism in the square chart, its Houses and the system of seven planets, all intelligent and adequate symbols ultimately indicative of the metaphysical mysteries of the Self and the Now. In truth the decline of astrology as an integral, sacred science began in ancient times, but in the modern revisions of Western astrology we see it reach such a point of decline that some of its most central symbols can be disturbed and profaned. In the modern world, of course, astrology is merely a fringe science and it is hardly a matter of cultural urgency that our astrologers have succumbed to significant errors of symbolism. But in the bigger scheme of things, and remembering that astrology is the most ancient and most sacred of traditional cosmological sciences, such errors are symptoms of more profound disturbances. And because astrological symbolism is so fundamental—so adequate—it is not surprising to find that the errors of symbolism into which modern astrology has fallen are not random or meaningless errors but rather errors that reveal the very nature of the modern deviation from traditional integrity. The pioneers of modern astrology had as their self-appointed task 'bringing astrology up to date' and indeed they recast Western astrology in a new image with a round chart instead of the square, a simplified House system with an emphasis on the Ascendant, 'bodily' aspects of longitude, nine planets instead of seven, and so on.[16] At every

15. The idea that the Tarot preserved an occult wisdom in the guise of a card game is an example.

16. It should be added here that many features of this modern astrology—such as the emphasis on the Ascendant—are reformulation or revivals of

instance, the changes they made constituted a fundamental realignment of astrological symbolism to accommodate a comprehensively horizontal world-view. In this respect, then, they did their task thoroughly. The instant popularity of the round chart and the sudden death of the square chart after thousands of years of consistent use are evidence that they hit upon a realignment at one with the profoundly anti-traditional and quickening *zeitgeist* of our times.

---

features of decadent ancient Roman astrology. The change of polarity from vertical to horizontal is already clearly evident in Roman astrological sources. It is around the beginning of the Christian era that the shift of emphasis from the solstitial to the equinoctial axis took place—so that the zodiac is counted as beginning at Aries, not Cancer—which is already a replacement of horizontal for vertical symbolism.

Astrology, Autochthony & Salvation

Not only in the plants that grow from the earth but also for animals that live upon it there is a cycle of bearing and barrenness for soul and body as often as the revolutions of their orbs come full circle, in brief courses for the short-lived and oppositely for the opposite. But the laws of prosperous birth or infertility for your race, the men you have bred to be rulers will not for all their wisdom ascertain by reasoning combined with sensation, but they will escape them, and there will be a time when they will beget children out of season.
Plato, *Republic* 546 A–B.

IN a system of astrology preserved for us in a short treatise by the late Byzantine writer Julius Firmicus Maternus, which system he attributes to ancient Egyptian sources, we are supplied with a symbolic horoscope or geniture of the world. This horoscope or geniture—which is only rarely mentioned subsequently in the western astrological tradition—is called the 'Mundi Thema'—Theme of the World—and, in this system, it is against the Mundi Thema that all particular genitures of individuals are to be compared and contrasted. Prognostications and diagnoses are made on the basis of the similarities and differences between the Mundi Thema and the particular geniture in question. The Mundi Thema is—symbolically and not literally, as Firmicus Maternus insists—a horoscope made for the moment of creation, the moment of the birth of the *kosmos* or for the pristine organization of cosmic forces. The births of individual souls are to be considered against it and the value of particular astrological configurations is determined in this way. We are supplied, therefore, with the primordial or archetypal horoscope and it forms the basis of all further considerations. The therapeutic correlative of this method is that each individual soul must, given the gifts with which they are born, attempt to bring their particular 'distorted' horoscope back into a realignment with the primal horoscope. In so doing they participate in or are reunited with the creative cosmic moment and lose, so to speak, their particular nativity and are instead born *with* the *kosmos, with* the world, *with* the Earth. In other terms, they reunify microcosm with macrocosm, healing the primal breach between the two. Although antiquity has not left sufficient remains to trace the most remote origins of this system even brief meditation upon the Mundi Thema reveals that it is indeed a primal crystallization of astrological symbolism and that this Byzantine writer has preserved a real key to the horoscopic arts. It is an astrology consonant with, and that no doubt was once part of, a much fuller cosmological understanding, some of the themes of which we will explore briefly here.

Since we are concerned with natal astrology as opposed to other types let us first consider the correlatives of this system to the plain physiology of birth, since the human body gives us the most primal order of symbols. In this case, the crucial moment in an individual nativity is the actual moment of *partum* in which mother and child are separated and the child emerges as a distinct entity, whether this is defined as the first breath of the child or the cutting of the umbilicus or some other exact juncture. By correlation, in this system, this event must be compared to the birth of the *kosmos* and so we observe that the emergence of the spherical head of the child from the birth canal of the mother is directly analogous to the birth of the cosmic sphere. The comparative roundness of the human skull betokens the spherical form of the *kosmos* itself, and the round human head is microcosmic of the *kosmos* as a whole, and since each individual nativity is to be compared to the nativity of the world, human birth is analogous to Creation and the emergence of the child's round skull from the darkness of the birth canal is therefore analogous to the emergence of the *kosmos* from primal Night. But as anyone who has attended a birth or is familiar with new-borns knows well, the process of birth invariably distorts and misshapes the human skull away from its pristine and geometrical inter-uterine roundness: the passage through the birth canal invariably elongates the skull and so in the passage from its life in the watery womb to its life in the airy realm the human being has its head undergo a fundamental morphological distortion. This is a mark of the particularization of individuality. While we once *all* shared the spherical head of the foetal state, our births have all distorted our heads in unique ways so that we all suffer a unique corruption from the ideal. In terms of the astrology we are considering, the Mundi Thema is the pristine ideal and the deviation of individual horoscopes from the Mundi Thema is a measure of the type and magnitude of the distortion of the spherical forms of the head at birth. This is exactly as Plato describes it in his cosmological treatise the *Timaeus*. He describes the spherical cosmic sphere then describes the spherical human head, drawing parallels between the motions of the stars and the very structures of human thought. Then, he says:

> the motions which are naturally akin to the divine principle within us are the thoughts and revolutions of the universe. These each man should follow, and correct the courses of the head which were corrupted at our birth, and by learning the harmonies and revolutions of the universe, should assimilate the thinking being to the thought, renewing his original nature, and having assimilated them should attain to that perfect life which the gods have set before mankind, both for the present and the future.

The system of Firmicus Maternus is nothing more than an astrological extension of this idea. It will be noted that the 'courses of the head' are 'corrupted at birth'. The spherical human head, which is a microcosmic reflection of the cosmic sphere is, in the travail of birth, invariably disaligned with its macrocosmic counterpart, and the horoscope is a record of the heavenly configurations at the exact moment that this disalignment occurred. It is as if the peculiar configurations of the heavens—representing that precise and unique moment in time and place—is imprinted on both the soul and the body of the child in a fundamental disalignment with the greater universe. As a cure for this predicament, Plato recommends that men study the 'harmonies and revolutions of the universe' in order to 'renew his original nature'. In this system of astrology the Mundi Thema is an image of 'original nature' and our geniture is an image of the type and degree of our deviation from that nature. Plato says we should 'assimilate the thinking being to the thought' and 'correct the courses of the head', bringing our own internal, skewed 'Circles of Sameness and Difference' (our nativity) into realignment with the 'Circles of Sameness and Difference' in the macrocosm. Thus will we enjoy the perfect life, both for the present and the future, which is to say we will attain what in other traditions is called *salvation*.

In terms of addressing a particular geniture and in the context of the practical symbolism of a horoscope, each configuration is compared to the pristine arrangement in the Mundi Thema and judged accordingly. In the Mundi Thema, for example, the planet Saturn is located in the (diurnal) zodiacal sign of Capricorn. If the planet Saturn is located in, say, Libra in the nativity of any given person, one judges according to the character of this relationship (or deviation from the ideal), in this case a deviation of ninety degrees. Similarly, if the Moon, which has its pristine abode in Cancer, is in Capricorn in any given nativity, it stands at 180 degrees or in opposition to the ideal, and so is judged accordingly. This single feature of the system Firmicus Maternus sketches also initiates two cornerstones of traditional astrological practice in the West, the doctrine of planetary

'domiciles' and the value given to 'aspects', though their fundamental relation to the Mundi Thema is generally forgotten. The
places given for the various planets in the Mundi Thema are
none other than those places traditionally known as 'domiciles'
or 'houses' or 'rulerships' whereby one planet is said to have its
'home' in a particular zodiacal sign. In the Mundi Thema the
planetary 'homes' are their places on the day of Creation, the
places where they were first placed in the heavens by the Creator. The method of determining relations between the actual
position of a planet and its 'home' in angular terms in this system is none other than the traditional method of determining
the value of 'aspects' which is still in regular use in Western
astrology today. By 'assimilation', as Plato calls it, we can overcome these deviations and slowly restore the planets to their
homes and our souls to their original state. We can never be *not*
born with Moon in Capricorn or Saturn in Libra if that is our
fate, but we can overcome these configurations and restore the
planets to their 'domiciles', which is to say to their pristine condition. This is the means to salvation. It is also the Pythagorean
doctrine of the harmony of the spheres, by which the microcosmic order is reattuned to the macrocosmic order by way of
their pristine unity. We might also transpose this to alchemy by
regarding the planets in their metallic incarnations (as indeed
Plato does in the *Timaeus*) and say that the task is to refine the
admixture of metals given at birth back to their pure, original
ores. In terms of physiology, one must restore the spherical
purity of the womb and so in that sense be 'born again'. In Judeo-
Christian and Islamic terms the Mundi Thema is representative
of the Edenic state, and the deviation of the planets from their
original homes is indicative of post-Edenic deviation. To restore
the planets of one's inner *kosmos* and its 'Circles of Sameness and
Difference' to their original stations, or to refine the planetary
metals to their pure ores, or to 'correct the courses in the head',
it is to restore the Edenic state, the dawn of the world, and to be
reborn *with* the world anew.

The Edenic comparison brings us to another point and to a
vital extension of this whole symbolism, namely that all that can

be said of the *kosmos* is interchangeable with the symbolism of the primordial man, Eden with Adam, and so we must not neglect to consider the Mundi Thema not only as the horoscope of the world but also—even though Firmicus Maternus does not elaborate on this—as the horoscope of the primordial being. In the Edenic state, of course, there is not yet a breach between the microcosmic and macrocosmic orders. The horoscope of Eden is the same as the horoscope of Adam. Adam is made of the very same soil as the Garden. If it is objected that Adam was created some time after the creation of the world, as such, in the Genesis account, the reply is that *all* events in Eden have the same horoscope until Adam and Eve are expelled from the Garden. This is for the simple reason that the paradisiacal Garden is in eternity beyond mundane time or is actually 'in' the eternal 'now' that is before, beyond and between time. To actualize this in oneself is salvation, for it is to find an unshakable 'safe' point outside the vicissitudes of the temporal order, including one's own existential encounter with the temporal order in death. In astrological terms, salvation is to cleave to and identify with the center-point of the horoscope, no longer tossed around in the to-ing and fro-ing of the periphery. And this is, at the same time, to realize within oneself the primordial man: to make as one's own the horoscope of Adam. To be saved a man must overcome or put aside—make irrelevant—his individual nativity and adopt the nativity (and spiritual anonymity) of Adam, *everyman*, who, in turn, has the same nativity as the very soil from which he was made.

A still fuller perspective opens up at this point. Here we encounter the mythology and doctrine of *autochthony*. To make this connection we merely need to appreciate that certain mythologies express the idea of being born *with* the Earth in the more concrete form of being born *from* the Earth, but the meaning is in both cases identical. In autochthony mythology, such as that of Adam, or that which we find among the ancient Greeks and the Athenians especially, primordial man was the product of a creation directly from the primal soils. In Greek mythology this is a characteristic of the fabled Golden Race: but whereas in

Adamic mythology the primal man was crafted from the soil, in Greek mythology they were plant-men, born from the soil. In either case, the symbolism here concerns the primal unity of macro- and microcosms: primordial man is still in harmony with the *kosmos* and the measure of this original, harmonious identity is expressed by saying that he was made from or came from the soil of the primordial earth. In Greek mythology, this is also expressed—noting the alchemical significances—by saying that the primordial men were *golden* in quality, being the very purest and best of souls, before the inevitable degeneration of the Ages which is usually represented by a decline away from autochthony into animal reproduction.

Transposed to the astrological framework, the great souls of the Golden Age are those primordial beings made at the creation who, therefore, *all* have the Mundi Thema—the Creation—as their geniture. The myths describe their excellence in terms of an aurumic symbolism and their conjunction with the cosmogonic moment and their primordiality is expressed in terms of being born from the soil. The fall into animal reproduction—where human beings begin to be born from one another and not from the soil—is the beginning of particular genitures. Our particular geniture is an image of our 'animal birth'. To 'assimilate' the Mundi Thema and to make the 'horoscope of Adam' one's own is, in this mythological view, to be reborn, to overcome one's 'animal birth', to restore the primordial aurumic quality of one's soul, to restore the unity of macro- and microcosms, and so to be of the *earthborn*. In astrological symbolism, this means overcoming one's individual horoscope and restoring each of the planetary powers to their original purity so that, in effect, the day of Creation becomes the day of one's birth and one rediscovers in oneself the primordial man. The distinctive Christian development of the themes of autochthony deserve special mention. Like Islam, Christianity

takes from Judaism the Adamic autochthony but adds to it Christ as the new and regenerated Adam. In this Christ appears as the primordial plant-man, firstly by being crucified (imposed) upon the Tree of Life which grows from the place where the skull—noting the physiological resonances of this detail—of Adam was buried like a seed, and secondly, and more obviously, in being resurrected from the grave in the Christian adaptation of ancient vegetative mythology. In the Christian scheme one must, like Christ, redeem and reattain the Adamic state, the autochthony expressed quite directly in the doctrine of the bodily resurrection from the grave.

Among these ideas we also encounter one of the secrets of the sacred calendars of the ancient world; namely that they embody a system of astrological autochthony. They do this by way of a parallel between the day of the New Year with the Day of Creation. In ancient Athens this involved the astrological calculation of the state wedding festivals. All of the young couples who had met in any given year were married at a single festival and in a single, collective rite. This was calculated to be nine months—the term of a pregnancy—before the New Year, which in Athens was marked by the Panathenaea, the great festival of Athena herself, also the festival marking the birth of the original Athenians, the autochthons, from the sacred Attic soil. The calendar was crafted, therefore, so that a high proportion of Athenian children would be born on or around the New Year. Anthropologists will explain that 'primitive' societies manipulated births in this fashion to ensure that children are born in seasons of plenty, and consistent with this in Athens the Panathenea was at midsummer, the fruiting season. But more significant, for our purposes, is the fact that it was New Year and hence the symbolism: *the New Year is to the solar cycle what the day of Creation is to cosmic cycle.* Children born at New Year in ancient Athens were presented with golden serpentine necklaces—noting the alchemical symbolism again—as an emblem of their autochthony; by being born at New Year they symbolically participate, by parallel, in the day of Creation and the birth of the primordial golden-souled autochthons. In this way—by 'seasonable marriage'—the

Athenians sought to prolong the original 'golden blood' of their mythic ancestors. As it happens, the Panathenaea—calculated by the new moon after the summer solstice, a lunar festival tied to the solar cycle—falls in the sign of Cancer, and this is exactly where Firmicus Maternus has situated the pivot of Creation in the Mundi Thema. The Western astrological tradition—following Roman models—is accustomed to thinking of the zodiac starting with Aries in Spring. Firmicus Maternus, using a much older arrangement, has the zodiac starting in Cancer in midsummer, which we note is the time of the Panathenaea in ancient Athens. In today's astrological terminology we would say that Athenian citizens tended to be Cancerians: their marriage festivals were organized so that as many children as possible were Cancerians. Why? Because Cancer marks the New Year and hence the Day of Creation, and children born at this time partake of the aurumic nobility of their earthborn ancestors.

This is a literal enactment of the astrology we find in the Mundi Thema, an astrological eugenics in which a society attempts to manipulate the time of birth of children to give them all similar horoscopes so as to impart to them a particular quality. In other Greek cities—in fact in many other ancient societies in general, including among 'primitive' tribes—the same effect was achieved by festivals or customs in which all the men or all the women would depart for a period of time and all return at an appointed time so that the women would all conceive together. Aside from anthropological explanations, these customs involve calculations of an astrological nature and are a device by which traditional societies attempt to prolong the original, aurumic essence of the Origin to which they characteristically cling. Traditional social and religious orders seek to cling to the state of primordial purity, resisting the inevitable decline of the Ages. One of the ways in which they do this is by the deliberate calculation of conceptions and births to try to give their children horoscopes that deviate as little as possible from the best of horoscopes, the horoscope of the Origin. It will be noted that this imparts a quality that is quite independent of blood-line, the usual measure of nobility.

As far as retaining something of the aurumic essence of the Origin is concerned, it is not of *whom* one is born but *when* one is born that is important. Aristocracy by blood-line is secondary but still reflected in the idea that the admixture of corrupt blood-lines will yield head-shapes that deviate more and more from the spherical ideal (and yield children who are less and less handsome). But the aristocracy imparted by 'seasonable marriage' as opposed to 'breeding' is the aristocracy of the soul of which Plato speaks in the *Republic,* the spiritual aristocracy of the philosopher-kings who rule, not by virtue of noble lineage, but by virtue of their inherent philosophical nature. The philosopher-kings are cultivated from children born into any class, independent of blood, who display the philosophic nature which, as Plato himself says, means any evidence of the persistence of the 'Golden Race' in the character of a child. The purpose of an astrological eugenics is to maximize the possibilities of such children arising. The qualities of the primordial race will persist in a child whose horoscope has been crafted to be as near to the ideal as possible, and moreover such children will have less deviation from the ideal to overcome and so will have spiritual and philosophical realization—realignment with the ideal—within their grasp and are equipped from birth, by nature, for the education and discipline that makes a philosopher-king.

In the conditions of the Latter Days, of course, conception and reproduction happen willy-nilly and children are born in all seasons and the horoscopes of men deviate further and further from the ideal and the rectification of the planets to their domiciles becomes more and more difficult. Men lose contact with the Source. Adam is exiled further and further from the soil from which he was made, in deeper and deeper alienation. The 'courses of the head' become hopelessly scrambled. The golden quality of the primordial order becomes increasingly rare. The task, in terms of the astrology we have been considering, remains the same, however: to compare our particular nativity to that of the Theme of the World and to attune the two. In this way we restore the Edenic unity of man and *kosmos* and occupy the eternal 'now'. The Mundi Thema is, in fact, 'contained within', so to speak, the primordial point that is at the center of any given horoscope and awaits realization. The particular configurations of a nativity are like so many shackles that bind us to time and space and to the illusion of our own egos. To be free of them all is to negate them, or to purify them, so that they return to their primordial stations which is, by a parallel symbolism, the same as all configurations 'collapsing' into the center where, in so far as they are still considered as separate powers, they are now in their Edenic and aurumic perfection. To achieve this is to mend our human alienation and to become, as a early Greek inscription put it, a true 'child of earth and starry heavens'. In the symbolism of both astrology and alchemy salvation is a return to the Edenic state by a realignment or reattunement of the microcosmic and macrocosmic orders. The system of astrology recorded by Firmicus Maternus, with the primordial horoscope, is clearly a key within this broader tradition and illuminates many of the most important but least understood aspects of ancient and esoteric astrology, exposing a fundamental vein of traditional symbolism that is rarely if ever explored in contemporary astrological thinking.

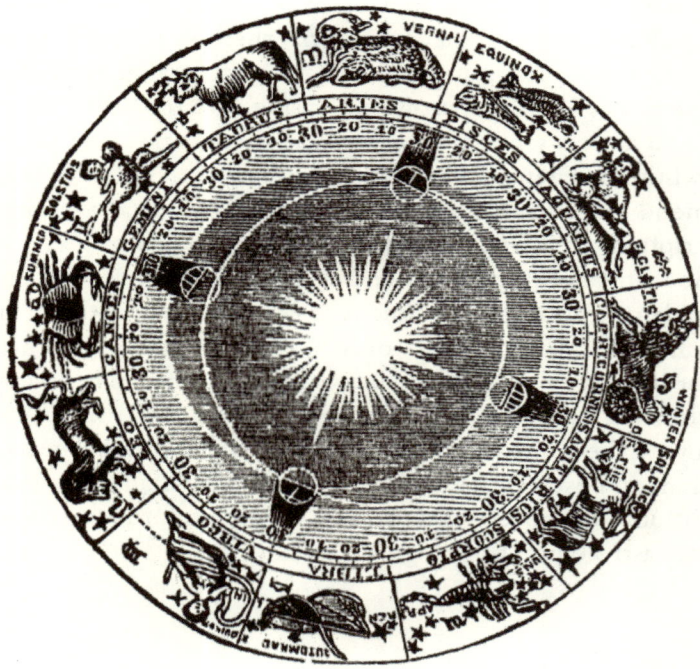

The Alchemical Symbolism of the Organs of Generation

When the work of bisection was complete it left each half with a desperate yearning for each other, and they ran together and flung their arms around each other's necks, and asked for nothing better than to be rolled into one.
Aristophanes, *Plato's Symposium*, 190e

THE human body is the most complete and primal of symbols and within the body the organs of generation are primary symbolic expressions, not only because sexual symbolism is primally creative and cosmogonic, but because, in a unique way, the genitals reiterate the symbolism of the whole organism in miniature and are an encapsulation of the whole symbolism *in toto*. The organs of generation, that is, are microcosmic of the human microcosm and in this have a symbolism that is comprehensive. This is suggested most strongly by the penis of the male which indeed seems like a 'little man' when erect, having a body and a head and imitating the uprightness of the human posture. There is something distinctly anthropomorphic or homunculean about the phallus. Plato's characterization of the genitals as 'self-willed creatures' having a mind and will of their own, independent of the being to whom they are attached, is a fuller expression of the same idea: the penis not only looks like a man but also has a mind and will like one, a localized autozooic creature. In the case of the female, the clitoris reiterates this, miniaturized yet again, and otherwise the internalness of the female organs describes the ensouled microcosmicness of man as opposed to his outward appearance, but again with a will separate from reason. In Plato's account of the reproductive organs it is remarkable that they are added to the primordial being as an afterthought, like ornaments added to an already completed form. They have an independent organization that repeats the organization of the whole: man-like ornaments added to a man. It is important to note, though—and this is what is unique about this order of symbolic expression—that neither the male nor the female organs contain the full symbolism but are complete only when considered together. It is actually the union of the two that fulfils the whole symbolic reiteration of the primal being.

Sulfur, Mercury and Sal

To explain this further, and as a means to illuminating an essential symbolism, we will resort to the use of the three key alchemical principles, Sulfur, Mercury and Sal, which principles, the alchemical tradition attests, are inherent throughout Creation as the cosmological correlatives of yet higher metaphysical hypostases. What is called Sulfur is an expansive, hot force and what is called Sal is its opposite, contracting and cold. Mercury, between these two poles, both expands and contracts and is characterized by oscillation and rhythm. The names given to these principles, of course, should not be confused with the profane chemical substances of the same names but should be understood as fundamental universal powers inherent in all things yet particularized more in some than in others. In terms of the geography of the globe we can think of the three principles being expressed in the contrast between the sulfuric tropics and the salific polar regions, with the mercurial temperate zones between. In the plant the hard roots that penetrate the cold of the earth are indicative of Sal forces while the flowering tops that seek the sun-filled air are Sulfuric in contrast. The leaves, with their cycles of respiration, are Mercurial. In traditional human anatomy and physiology this same threefold schema is extrapolated to the human form. The Sal pole, in this case, is found in the head, and the Sulfuric pole is found in the flora of the digestive tract and the flowering of the reproductive organs. The Mercurial forces are found in the organs of the thoracic cavity, the heart and lungs, characterized by rhythm. This organization, it is important to note, is exactly inverse to that of the plant. In the human, the head is the 'root' that is rooted to heaven rather than to the earth, and the flowering processes are in the lower organization, not the higher. This is most clear in the female, in this instance, both in her long hair—hair being like roots from the head—and in the distinctly floral appearance of the vulva and external genitalia. In the male the phallus is mushroom-like, and the same symbolism prevails, except that one must know that mushrooms are actually the flowering organs of a fungus in order to appreciate it. It is also clear if we consider the skeletal form of the human being: there we find three separate cavities, abdominal, thoracic and cranial, three interiorized spaces, that invert the functional organization of the plant.

The threefold organization of the whole body is reiterated by reflection in the various parts of the body. The isomorphic reiteration of the whole in its parts is an inherent principle of the human form as it is traditionally understood: the over-all threefold pattern is found reiterated everywhere. Most obviously, perhaps, the finger or digit is, like the penis, a 'little man'—as a finger puppet would remind us—but (unlike the penis) has an explicit threefold organization of bones and joints; the threefold organization of the whole organism is reflected in the structure of the ligatures. The same threefoldness is then repeated in the structure of the hand and arm, and so on. This principle of isomorphic repetition forms the background to many diagnostic and therapeutic techniques found in traditional medicine where, say, the ball, arch and heel of the foot reflect the whole body and so-called 'foot reflexology' works by virtue of this correspondence of the part to the whole. In sexual psychology, of course, it also forms the background to various types of fetishism which obsessively identify one part with the whole. The scheme of the three alchemical principles seeks to delineate the patterns of internal repetition that permit such identifications and, moreover, this extends beyond man to the macrocosm where the same patterns are found externalized. In truth, the three principles are expressions of three creative metaphysical hypostases, the source, cause and end of all things. The pattern is repeated throughout creation because all things share the same transcendent origin which reveals itself everywhere in this threefold manner—expansion, oscillation, contraction. We find the pattern in the macrocosm—essentially Heaven, Sky and Earth—while the human body as a whole is the fullest reiteration of the pattern in microcosm; but then, we must understand, the pattern is repeated within the microcosm itself over and over at every level forming the matrix of an esoteric human anatomy.

The Face

The fullest and most obvious reiteration of the over-all threefold pattern within the human microcosm is in the face. Here we find the brow and eyes corresponding to the 'head', the nose to the rhythmic organs and the mouth to the lower organs, both the digestive and reproductive. The eyes and brow are Sal, the nose Mercury and the mouth Sulfur. Again we can see this last correspondence most obviously in the female form where there is a direct correspondence between mouth and vagina made obvious by the fact that the vagina has lips. The face reiterates or mirrors the whole and thus can represent the whole and in practice erotic life is often played out by facial signals: lipstick refers to aroused vaginal lips, the moustache to pubic hair, and so on, as well as the kiss being an image of the sexual act, 'microcosmically', so to speak, a lovemaking between two faces. When lovers' mouths are joined in a kiss, it is symbolic of the greater union of genitalia because the face encapsulates the whole body and in this encapsulation the mouth corresponds to the genitals, and furthermore the experience of breath, eyes, taste that accompanies the kiss is an image of the totality of the experience of actual sexual union. Needless to say the tongue here takes on a phallic function and saliva becomes analogous to the sexual fluids. It should be observed, however, that, despite this relative completeness of the reiteration in the face—such that the kiss may be a reiteration of full sexual congress—the face is not, by resemblance, man-like, as is the penis or the finger. While the face repeats the threefold organization of the whole being in a particularly complete way, it does not *look like* the whole being. Instead, it is characteristically round—a round face on a spherical head. This signals an important principle: *the fuller the reiteration of the whole the more the isomorphism takes round and spherical forms*. This is because the primal man is—symbolically—spheroid, reflecting microcosmically the sphere of the *kosmos*. Considered in its totality the macrocosm is spherical. The spherical skull of man betokens the fullness of man's image of the macrocosm and the perfection of the primal man—the complete microcosm—is expressed by the image of a perfectly spherical being. Primal man, we might say, is, by this order of symbolism, *all head*—an idea that is suggested in the morphology of a new born human child with its disproportionately large head and an underdeveloped body that is seemingly but an appendage to it, like a cotyledon sprouting from a seed. Thus where we find the fullest reiteration of the whole threefold pattern, namely in the face, we find it in the context of round and spheroid forms, framed by the roundness of the face and located in the spherical head. We might say that wherever we find the three alchemical forces in their primal relationship, there we will tend to see rounded forms. The dome of the penis and the face of the thumb are small examples, accompanying an upright man-like appearance in both cases, but the face is the prime example in human anatomy. The face is not man-like by resemblance but *in organization*, and the extent of this manifests in roundness—an image of the principle, not a 'naturalistic' or homunculean imitation of the creature. Naturalistic imitation or simple anthropomorphism—by which the penis appears man-like—is a reiteration of a lesser order. The more fundamental and complete the reiteration the more we find round and spherical, not literally anthropomorphic, forms.

It is remarkable though that this same pattern of threefoldness and that symbolism does not apply to the genitalia of male or female considered on their own, but only when they are considered in conjunction. Despite appearing man-like and inviting such a comparison the phallus is not—as a finger is—tripartite like a man. Nor is the clitoris. While other parts of the body and especially the face of both man or woman reiterate the threefold whole, the organs of generation—which are distinct in man and woman—on their own, do not. The penis is a 'little man', but in fact it is only the *head* that is developed to mark this, although even then its incompleteness is signaled by its single eye. Similarly, although the vagina is like a mouth, it has—so to speak—been fitted sideways. There is something defective about these anthropomorphic reiterations. The penis is blind. The vagina is dumb. There is something comic or parodic in each case, again like awkward and ill-fitting afterthoughts to the whole human design. Neither are part of any obvious threefold organization in themselves unless we group the penis with the testes and the clitoris with the ovaries. Instead, it is necessary to appreciate that the male and female organs are each incomplete without their opposite and so each will only give us part of the whole symbolism. The symbolism becomes clear when these complimentary parts of the human anatomy are considered together as a pair. The face and other parts of the human form will present a more or less full symbolism in themselves, but the genital organs need to be considered in unison.

In this case, the penis is the Salific pole of the pair and the vagina and female organs the Sulfuric pole. The hardness, stiffness and external character of the male organ indicates a preponderance of the Salific force. In contrast, the soft, moist internalness of the female organs, as well as their odors, betokens a concentration of the alchemical Sulfur. The phallus is a symbol of Sal. The yoni is a symbol of Sulfur. We then find the intermediate force, Mercury, in the union of phallus and yoni, namely in the rhythmic movements of the sexual act. Schematically:

The hardening of the penis in the male = Sal.

The opening of the vulva in the female = Sulfur.

The rhythmic insertion of the penis into the vagina = Mercury.

It follows then, returning to the analogies we pursued earlier, that the penis is the 'head' of this organization, the vagina the 'belly' and their union is the rhythmic system (that in the body as a whole takes the form of heart and lungs, noting here the acute relation of breath and pulse to sexual arousal). If we are looking for the microcosmic reiteration of the threefold organization of the whole body according to the three alchemical principles in the human organs of generation we must consider them together, not separately. Together, they form the tripartite, microcosmic whole. The penis in itself—though it looks like a 'little man' and betokens this symbolism by appearance— is in fact a (one-eyed) concentration of Sal forces—its anthropomorphism is mainly in the head—just as the female organs are so sulfuric that the salific clitoris has been retarded to a miniscule size. When the (salific) phallus joins with the (sulfuric) yoni in the (mercurial) rhythms of love, however, the whole microcosmic symbolism is complete. Thus is an image of the primordial man formed. There is no obvious Sal-Mercury-Sulfur delineation to be made of the penis or the female organs in themselves: rather the penis is a microcosmic concentration of Sal (the most Sal organ in human physiology) and the *cunnus* of Sulfur (the most Sulfuric organ in human physiology) and it is only when considered together that they form the whole symbolism and repeat the full, threefold organization of the whole being. Alone they are—as it were—deformed by an unbalanced concentration of forces; together they form an alchemical whole.

Primal Sphere

Although it might not seem obvious, given the usual tangle of arms and legs, the male and female bodies united—at least in the uniquely human face-to-face embrace—form a suggestive rounded and spheroid shape, consistent with the symbolism outlined above. Shakespeare's Iago expressed a puritan horror at the 'beast with two backs', but the comic Aristophanes understood better when he compared male and female united as like a bouncing 'ball'. In the tale Aristophanes tells in Plato's *Symposium*—in which some of the symbolism we are considering is explicit—the primordial beings were sexually undifferentiated and rolled about like balls or spheres. Amongst other things this alludes to the suggestive two-backed shape made by lovers in the face-to-face embrace, seeming as if the two torsos constituted a single ball-shaped form. More exactly, in human physiology, it alludes to the distinct sphere formed by the two half-spherical buttocks of the united lovers. In male and female the buttocks, seen from the side, form a half-moon in each case—indeed they have a soli-lunar symbolism. When male and female come together in genital embrace the buttocks of the lovers together shape a sphere. This is sometimes very apparent in the characteristic movements of sex: the lovers move as if they are both involved in rotating and counter-rotating a sphere made by the connection of their lower regions and outlined by their buttocks. The sphere here denotes wholeness, totality, primordiality. While the threefold organization of the whole organism is repeated in the face, the finger, the foot and throughout the human body, in the genital organs of either male or female alone we find only one pole each: the threefold pattern only occurs in their union and *then* do they reiterate the organization of the whole man. Moreover, they do so in an especially complete and fundamental way and so, as we said at the outset, encapsulate the whole symbolism comprehensively. The threefoldness they reiterate is that of the primordial—symbolically spherical—human form. The image of this sphere is made by the joining together of the two half-moons of the buttocks. Lovers form one body, as the love songs say, but it is *the body of the spherical primal man.*

Astronomical Symbols

Spheroidal symbolism, of course, introduces astronomical cor-relatives, and here we touch a more profound level of the same doctrines. The phallus, especially its circular head, is solar while the vertical curved lips of the vagina are suggestive of the sliver of the New Moon. Here the Sun is a Salific concentration (the diurnal concentration of the stars) and the Moon sulfuric, but a more complete symbolism envisages the vagina as the terrestrial 'cave' of the sub-lunary realm, in which case the clitoris is lunar since the sub-lunary cavern is literally 'below the Moon': in this the Moon and its monthly rhythms is mercurial and the Earth the feminine, sulfuric pole. It will be noticed, though, that this symbolism can be reversed, for the hot light-emitting Sun can quite naturally be understood as Sulfuric and the hard, mineral Earth as Sal. This interchangeability is significant. In human physiology it is the order of symbolism that prevails in the event of orgasm, for here we witness a sudden reversal of poles. The male orgasm involves the sudden change of Sal into Sulfur and the female orgasm the opposite. In the case of the male, the hard, concentratedness of the phallus suddenly gives way to the sulfuric, expansive ejaculation which is followed by a (sulfuric) flaccidness. The female orgasm involves the same change of polarity. The expanding openness of the sex organs suddenly gives way to contractions—the womb contracts and the sex organs 'suck' inwards instead of expanding outwards in open-ness and invitation. In cosmological terms, orgasm corresponds to those critical inversions in cosmic history when one cycle changes to another. In mythology, for example, at the end of one cosmic age the sun, we are told, will suddenly rise in the west and set in the east and the whole of time will be reversed. The sphere spins the other way. The throes of sexual climax are characterized by sudden inversions in the same manner and by the same chain of associations reiterate the cosmic climaxes in miniature. The conjunction of Sun and Moon—or Sun, Moon and Earth—in eclipse is also a cosmic climax symbolic of the greater cycle *and of sexual climax*: it reverses, momentarily, the primal opposites day and night. Orgasm is equivalent to eclipse. The anguish of climax is like the sudden darkness in which the whole order of night and day—and all other opposites—are overthrown and, in the most sublime paradox of which the human creature is capable, the lovers are flung into holy Night *yet waking*. When Sun and Moon come together in eclipse, night and day reverse, just as at the end of a total cosmic cycle the whole order is reversed. So too when man and woman join together, the climax of their union is a sudden reversal of cosmic polarity.

This is most obviously the experience of the male who, at ejaculation, finds the rhythm of his thrusts broken by what seems an irresistible counterforce, as if the cogs of some inner wheel had suddenly been engaged in reverse. Consistent with this eclipse symbolism, the penis in this case becomes serpentine in character and the female part becomes a cave with access to the depths of the Earth. In the eclipse the celestial dragon devours the Sun and carries the solar, aurumic treasure into its subterranean lair. Thus the penis deposits semen—solar in essence, the dragon's gold—into the womb—the dragon's cave, and by extension, the alchemist's *athanor*. To complete this astronomical symbolism, it should be noted that the female pubis formed by the inner thighs and the inverted triangle of pubic hair makes the cross of the tropics and thus the vaginal entrance (or in fact the clitoris above it) marks the node where these tropics meet, the equinoctial points. The head of the phallus, as seen when looking down upon it, is the solar disk, that hurries to this junction. And we cannot mention the serpentine symbolism of the phallus without also commenting that the foreskin represents the skin that is shed by a snake in its vernal regeneration, and this in turn represents the shedding of Ages in the greater cosmic cycle. Herein is the true significance of circumcision. It is an emblem of rebirth. The circumcised penis is like the regenerated snake of spring, freed of its old skin. It is thus also a mark of the regenerated Golden Race that is reborn at the turn of the great cycle.

Modes of Coition

As far as actual modes of coition are concerned, the principle that the human order reflects the macrocosmic order once more holds sway, and the first thing to note is that the many forms human lovers can make in their lovemaking exhausts and exceeds all the modes found in the animal kingdom. Other animals are restricted in their mode of coition: dogs do it like dogs and birds like birds and rabbits like rabbits, but human lovers can enjoy all of these modes, and more. Thus in some ancient tantric manuals various sexual positions are known as the 'Lion' or the 'Scorpion' or the 'Gazelle' and so on: adepts in the arts of love have mastered the ways of all creatures, fierce and tender, and the multiplicity and zoomorphic plasticity of the possibilities betoken the completeness and dignity of the human form. Face-to-face coition, however, is, as mentioned above, uniquely human and what is known as the 'missionary' position does have an essential symbolism that makes it archetypal, at least among face-to-face positions, for the woman is the Earth and the man the Sky and they enact the cosmogonic and primordial nuptial of the two in an especially adequate way; in the same embrace, their respective faces become Moon and Sun. The symbolism of the reverse position with the female on top is not diabolical, as the puritanical advocacy of the missionary position implied, but of another order: here the arch of the female body made by her movements symbolizes the night Sky (in the manner of Nut in Egyptian mythology) and the male is the supine Earth below. In regard to the eclipse and dragon symbolism mentioned above, we here see the vagina as the sun-eater. The association of the female breasts with the starry heavens signaled etymologically in *galaxy* = *galactic* = *milk* is relevant in this case too. This position has a stellar, lunar nocturnal symbolism compared to the solar, diurnal symbolism of the missionary style. The female on top position is also liable to emphasize the axial nature of the solar phallus and the rhythmic—which is to say cyclic—movement up and down that axis. The natural angle of the phallus to the body in this position is not perpendicular, though, but is erect at an angle that is again suggestive of the inclination of the tropics and the ecliptic. Other positions, such as the so-called 'doggy' style—probably the most recognized style of zoömorphism in human erotic play—turn this axial symbolism sideways such that the phallus becomes like the axis in a wheel—formed by the rounded outline of the female buttocks again, now seen from behind, not from the side—and the appropriate metaphors found in traditional sources concern both the wheel and the chariot. Obviously, in some positions the male takes the active role and in some the female is active and the male passive. In general, either diurnal or nocturnal symbolisms will prevail in any given case.

Concerning anal sex, whether between man and woman or man and man, we need to observe that the genital organs are located in the same vicinity as the vents of bodily elimination, and in the case of the penis we find an organ that serves as both. It may seem incongruous that these organs should be located together on the body: in fact it is not just incongruous but actually paradoxical, and expresses the paradox that, in cyclic terms, *the end is the beginning*. The significance of this conjunction is simply that here we witness the anatomical expression of the idea of the beginning (generation) and end (elimination) of a cycle in conjunction. The spherical form of the primal being has left its imprint on man. The equator or central meridian of the primal man runs vertically, starting and finishing in the perineum. The clitoris marks the beginning in females and the penis marks the same beginning in males: in both the anus marks the end. In other respects the anus is chthonic or indeed—as per taboo—symbolic of the underworld and the realm of the dead, but all anal erotic play will involve a symbolism concerning the conjunction of beginning and end. In oral sexual contact, fellatio or cunnulingus, or both together, a similar idea is expressed more directly in the classical alchemical symbolism of *ouroboros*, the serpent eating its tale. It will be noticed that while genital intercourse has a *spherical* symbolism that concerns only one cycle both the anal and the oral modes of intercourse have a more *circular* symbolism that expresses *cycle* per se. Lovers in the 'sixtynine' position form a circle, not the spherical 'beast with two backs'. The difference is that genital intercourse is more polarized and axial in that it can only occur between man and woman, while anal and oral forms of sex have an inherent ambiguity in this respect even when enjoyed by man and woman. In their essential symbolism oral and anal forms of sex do not evoke the spherical primal being but rather the serpentine symbolism of the cycle, a two-dimensional as opposed to three-dimensional morphological expression. This makes them of a less primal character but they are not without a legitimate symbolism.

Sexual Fluids

The sexual fluids that issue from the genital organs have a symbolism that is consonant with all that has been sketched above and that further underlines a number of important points. The floral appearance of the female organs, for instance, invite a comparison between the female fluids and nectar and the characterization of the fluids as 'juices' underlines the idea that the genitals are the flowering and fruiting organs of the body. In the case of the female part this is explicit in some languages where, for example, it may be given the same word as a 'fig' or at least is called a 'fig' or some other fruit colloquially. The parallel is unmistakable in the case of semen which is very much like the white exudate or sap of certain plants including the most common of grasses and the pungent smell of which is often identical to that of many flowering herbs, shrubs and trees. What needs to be noticed here is that while human sexuality is routinely described as being part of our 'animal' nature, much of the symbolism of the organs of generation, including their fluids, is taken from the vegetative world—indeed, the organs of generation are the most *plant-like* in the human body. The reason for this is that the primal being, as ancient traditions attest, was a plant-man—a spherical 'seed' specifically—and was prior to the sexual differentiation of animal reproduction. Before the genitals were attached (like afterthoughts) to Plato's primal being, the primal being was a plant-man generated, in myth, directly from the soil of the Earth—which is to say, in metaphysics, the Ground of Being. The human genitalia reflect this primal state and even though they are the vehicles for animal reproduction—itself a consequence of a falling away from the autochthonous primal condition—of all organs in the body they retain and reiterate the morphology of the earlier stage.

By extension, we also find an accompanying *insect* symbolism because, in traditional cosmological understandings, the insect world is merely an extension of the plant realm, the insect being a flower or fruit that has become liberated from the plant into the air, so to speak. The insect world is inseparable from the sexual organization of the plant realm. Here we see the penis as the insect and the female organs as the flower. And among insects, it is the *honey bee* that conforms most fully to this symbolism, having the same solar associations as the phallus and reminding us—apart from the idea of the penis as the bee's sting that punctures the female to inject venom—that semen is very often described as 'honey'. This association is universal and persistent despite semen's white (not golden) color and salty (not sweet) taste. The higher solar symbolism of semen is essentially that of light. The sun too is white but commonly described as golden: we find the same transpositions regarding semen. The saltiness and smell of semen—also its *whiteness* from another viewpoint— betokens the concentrated alchemical salific force just as the tastes and odors of the female 'juices' signify a concentration of the sulfuric principle. It might be objected here that semen— and the phallus too for that matter—is characteristically *hot*, not cold like the salific pole to which we have ascribed it. Strange to relate, the heat of semen, and the phallus, is quite properly called a *cold heat*, admitting the paradox that there is a *cold heat* associated with the alchemical Sal. We can understand this by referring to the honey bee. While this eminently solar creature

seems quintessentially *hot* by temperament, it is in fact, and paradoxically, *cold blooded* like a reptile. This overlaps with serpent symbolism where snakes can be naturally either hot or cold in character—hot in their speed, their venom, the pain of their bite and often in their color—but, of course, (like a plant, it should be noted) having no inner heat of their own. The penis is, we might say, cold blooded like bee and snake (and plant); it hibernates and is aroused again with the Sun, although not over *winter*, but (by parallel) over *night*. This is the sense in which it's alchemical nature is cold and Salific.[17]

Insect and Snake

Quite apart from modes of coition that directly mimic those employed by animals, much human love-play, especially in the specifics of genital encounter rather than in forms made by the whole body, mimics plant/insect interaction and, especially where it is phallically centered, the character and modes of snake. In cunnulingus, to use a graphic example, the male is the honey bee lapping at the nectar of the flower. In some types of vigorous coition, especially as climax nears, the male seems like a bee frantically injecting the female with its sting, and just as the bee dies when it releases its sting, so the male collapses into a death-like sleep after ejaculation. The phallus is most obviously snake-like in masturbation, in which case it rears up and spits its cold-hot venom. 'Venom' here, as with the bee, does not signify a *poison* so much as a concentration and *potency* and the healing qualities associated with bee venom and snake venom (and honey) in traditional and alchemical medicine are as active here as any

17. It is also said to 'quench' the female fire.

other associations, including parallels between the throbbing, swollen redness of a bee-sting and the head of the aroused penis and between the delirium, spasms and fever that is typically induced by snake toxins and the fevered frenzy of copulation.[18]

Hair

Regarding the *hairiness* of the external sexual organs, the thing to note is that in the human form in general hairiness has become particularized to these organs and that the rest of the body, other than the head, is comparatively naked. This nakedness in fact serves the genital organs; it makes the human skin especially sensitive and endows it with a heightened eroticism to which the genitals respond. What is remarkable is not that the genitals are hairy but that the rest of the body is not and that this fact gives a special sensuality and totalness to human eroticism such that the skin is like an extended erotic organ. This is concomitant with the microcosmicness of the genitalia; conversely, the whole human body, in its hairlessness, partakes of the erotic sensitivity that is proper to the genitals. If the penis is a 'little man', there is a sense in which a man is a 'big penis' and it is in fact signaled by the comparative hairlessness of the body which allows genital sensitivity to extend throughout the whole external creature. The human body is made for love in its whole organization. Human sexuality is particularly total and is not localized to a genital experience.

Conclusion

In summation, the order of symbolism explored in this present article is, as we have insisted, primordial and fundamental in

18. In homeopathy many of the snake toxins such as Lachesis and Naja display a bizarre sexual psycho-pathology with such symptoms as the manic urge to expose the genitals, compulsive lewdness, and so on.

nature, and for this reason it is also especially sacred. Not only does it give a comprehensive reiteration of primal symbolic patterns, it does so in a unique way, namely by polarization between man and woman, sal and sulfur, which when brought together form the completed alchemical triunity of forces in their primal relationship. The organs of generation, we might say, are more primitive and archetypal, and invoke more basic cosmological correlations in a more primary way, than the rest of the human body, even though, as we have said, they are 'added last' and represent, in the full scheme of things, and despite their vegetative primitiveness, a fall into animality compared to the primal being. We have sketched this symbolism from the purely alchemical view point, but in practise this and other sexual symbolisms are subject to the restraints and parameters set by the spiritual and moral climates shaped by traditional religions which may develop or suppress aspects of this type of symbolism according to their particular needs. This symbolism is integral, however, and in all but the most monastic and anti-erotic environments we will find cultural expressions of its themes, if not a properly constituted 'tantra' or erotic method of attainment. That sexual union is an alchemy with an adequacy and completeness of symbolism for it to act as a vehicle for spiritual realization is admitted in many traditions.

Sexual union has the potential to transform the whole being, just as it visibly transforms the face of a lover, sometimes as if to reveal a whole other person. The ecstasy of sexual climax is a reflection of Divine Bliss and indeed the fullest reflection of it that is within the experience of most people. The great secret of 'tantric' alchemy, however, is not only in the climax itself—nor even in its postponement, which is merely a physical device to heighten climax when it can be postponed no longer—but in its aftermath, namely the profound, metaphysical *sleep* that follows. It is not by any means an accident of nature that intercourse is typically followed by the urge to sleep and that this sleep is often of a particularly profound and regenerative variety. It is in that sleep that the symbolism of the act is realized cosmically. The waking consciousness is solar and phallic and the dark realm of sleep is the vagina, here not chthonic but rather celestial Night, with the vulva as the gates to her palace. The palace of Night is the Sphere of Being. Just as the solar phallus penetrates to the very center of the sphere made by the union of the lovers—outlined by the buttocks—so the solar consciousness may penetrate to the very center of sleep, the deepest center of deepest sleep and there find what the alchemical tradition allegorizes as the Fountain of Youth. In climax the existing order crumbles. The cycle is reversed. Phallus and vagina exchange polarities. For a split second a chasm opens and the precious center of Night is revealed. Just the glimpse of this is the whole source of the lovers' bliss. And as order restores itself the center of Night, having been exposed, drags the solar consciousness back into the darkness of sleep, irresistibly, precisely corresponding to the orgasmic, centerward suction of the vagina drawing in the solar phallus. The organs of generation—strange tragicomic actors though they seem—enact a cosmic and ultimately metaphysical drama.

Evolutionism & Traditional Cosmology

WHILE it is routine for writers from a traditionalist or perennialist perspective to compose condemnations of Darwinism and to expose what Titus Burkhardt, in a celebrated article, called 'The Transformist illusion'[1] it is rarely acknowledged that the evolutionist doctrine is, in part at least, a corruption of a traditional doctrine. *There is nothing new under the Sun,* as the Preacher sayeth, and the novelties and 'discoveries' of modernity are either misconstructions or negations of traditional ideas and forms. Darwinism is no exception. We find it prefigured in traditional accounts by which the human microcosm reflects—and is 'coagulated' or 'extracted' or 'condensed' from—the macrocosm. No less a representative of Tradition than the Muslim sage Rumi gives a famous example of an 'evolutionary' perspective:

> I was a stone and I died as a stone and was born a plant. I died as a plant and was born an animal. Later I died as an animal and I was born as a man.

Those who attempt to marry modern science with traditional wisdom very often quote this passage from Rumi as a way of saying that the sages of old had an intuitive knowledge of truths that Darwin made concrete and scientific. In part—but only in part—they are right to do so. Here Rumi reiterates a sequence of

---

1. Taken from the title of the book *The Transformist Illusion* by Douglas Dewar, Sophia Perennis, 2005

states that at least resembles the Darwinian account of man, for by Darwin too man was once stone, and flower, and so on, in a progressive sequence. Certainly, Rumi does not suggest 'natural selection' as the device by which he 'dies' from one form after another, but he nevertheless understands the human state as the fulfillment of a sequence of creatures, each more complex and 'evolved' than the previous. He was at first a stone—inert matter. Then vegetable. Then animal. And at last human.

But the modernists are wrong to suppose that Rumi is entirely at one with Darwin and that modern, quantitative science is the fulfillment of ancient wisdom traditions. Rather, we must understand Rumi in the context of the traditional cosmological sciences and, in the case of this passage, realize that Rumi is giving expression to the cornerstone of traditional cosmological thought, the microcosm/macrocosm doctrine. He is describing the condensations of the macrocosm into the human microcosm which—by all Traditional accounts, and by definition—contains, in essence or in tincture, the whole of the macrocosm. Man has a nature that is stone, and vegetable, and animal, which is testimony of his 'extraction' from the macrocosm. The profane doctrines of evolutionism bear a *resemblance* to this in so far as they propose that the human being has emerged and is constituted from 'the environment'. Traditional sources more often describe this in terms of an 'involution', since the microcosm is an 'interiorization', and without the progressive and 'evolutionary' sequence used by Rumi, but the crude notion that man has emerged and is constituted from his external world need not be ruled anti-traditional in itself, provided we understand that the modern doctrine is, all the same, a hopelessly limited and partial view—of both man and the universe.

The traditional doctrine—in an admittedly simplified and incomplete rendering—can be presented in the following few points:

1. A Metacosmic Principle—Pure Subject in contemplation of Its own Object, Identity, at once Unique and Infinite.

2. The macrocosm is an 'exteriorization' of the Principle (as Object) through the microcosm.

3. The microcosm is—at the same time—an 'interiorization' of the Principle (as Subject) through the macrocosm.

This is leaving aside any account of 'patterns' or 'forms' or 'archetypes' or any further distinctions (hypostases) that reside in and are manifested from the Principle. It is enough to say that there is a macrocosmic order and a microcosmic order and these are complementary expressions of the same Principle which fact is the basis of their mutual reflection. That is, they reflect each other as well as (and because of) reflecting the Principle—and this because of the very nature of the Principle Itself. Within the non-manifest Principle there is neither an inside nor an outside, subject or object; it is beyond but also the root of these dualities—which dualities, therefore, are not final. Taoism, which embodies and preserves the ancient 'alchemical' perspective more explicitly than other traditions (and is in some respects, we might say, the least theological and most cosmological of religions) depicts this arrangement in the classical yin-yang symbol where the yin is contained in the yang, and vice versa, and both have identity in a non-manifest Principle (The Tao). In the occident, the caduceus of Hermes, and other symbols with intertwining serpents or dragons, represent aspects of the same thing. There is an unfolding and an infolding at the cosmological level, but no movement at all at the level of Principle.

For our present purposes the thing to note is that point 3. allows for the idea that man is an extraction of the *kosmos* and a reorganization of macrocosmic elements. The organization of man reflects the organization of the *kosmos*, and this because he has been constituted from the *kosmos*, or rather *from the Principle through the kosmos*, which distinction is all-important. It will be seen that evolutionism is a specific misconstruction of point 3. at the level of this distinction and is an overall misconstruction since it is ignorant of points 1 and 2. But in the first instance there is nothing altogether illegitimate about the notion that man is constructed from and has within him the fire, air, water and earth that are the constituent elements of his abode. It is not even necessary to make the proviso that it is only his material frame that is so constituted, for even his 'consciousness' may be taken as an internalization of the light that illuminates his abode and so his 'consciousness' is in that sense derivative from his 'environment'. His waking and sleeping are an internalization of Sun and Moon.

In such a perspective it is entirely possible to conceive of man as the culmination of a succession of animal forms, each more completely internalized than the previous. It is possible, then, to conceive of this internalization as the key to 'survival of the fittest'—fittedness being a measure of macrocosmic involution— and we may even hypothesize, with Darwin, that chance mutation is the propelling device. That is, life 'evolves' from inert matter by chance mutations, and those mutations which give rise to internalized forms survive in so far as internalized faculties— because they are reflections—enable a creature to respond successfully to its external circumstance. At length, a creature (*homo sapiens*) 'evolves' that is a virtual reflection of the whole *kosmos*. It is possible to conceive of this as having taken place gradually with other less successfully microcosmic forms appearing along the way, the gradual and linear trajectory of the process being a consequence of the temporal arena in which it occurs.

To adapt a traditional symbolism to this, the living entity and the universe that is its environment are as *mirrors* to each other, and the fossil record appears as a process of bringing the mirrors into alignment or into focus. The mirrors move by 'chance' forces, let us say, and are sometimes near to focus and at other times wide of focus, until—by 'chance' forces, let us say—they hit an alignment that finally yields a true or near-to-true reflection, namely the human form. And let us also say, conceding further to Darwin, that 'focus', in this analogy, is the key to creaturely survival. Man won the race because the human form is the better focus between the two mirrors; this is what it means when we say he is adaptable; his internal resources correspond best to the requirements set by his external world. But, let us remember, there is no 'chance' at the level of Principle and the appearance of 'chance' at the cosmological level is an illusion. (The Greeks more correctly called it 'Necessity'.)

And, more importantly, where these mirrors—entity and world—reach focus they reveal the Principle that is responsible for their correspondence and the basis of the 'mirroring'. *The true reflection reveals the principle of reflection.* This is the point at which the subject/object duality is resolved; what is inside is outside and what is outside is inside. The metacosmic Principle is beyond subject-object complementarism and resides in its own Isness, having no complementary opposite, Pure Subject in eternal self-contemplation, its own Object.

There is no need to say anything further about this Principle, for we are considering its cosmological function and not its metaphysical content, and it is especially unnecessary at this level to introduce theological subtleties: the point is that man is not only the summation of the macrocosm—the subject that answers to its object—but he also embodies the Principle that resolves and transcends subject/object and so is *a spiritual or transcendent being.* This means, precisely, that he is capable of grounding his being in the Principle and make the 'point of view' of the Principle his own, so to speak. Darwinism, along with modern thinking generally, is guilty of the most appalling underestimation of the ontological range of man, but so far as they go the general propositions of evolutionism may not be entirely deviant.

Finally, we should note that traditional accounts tend to give priority to point 2 because Subject is logically prior to Object. If it is true to say that man is an 'extraction' of the *kosmos* it is nevertheless more true to say the *kosmos* is a 'casting off' or 'excrescence' or 'evaporation' or 'filtration' of or 'projection' from man—or rather through man (and, if you like, from Man, i.e. Primordial Man, Adam Kadmon, Purusha.) Darwinism, of course, has no notion of this and so is hopelessly partial in its perspective and for that reason destructive to the wisdom traditions of the ages. It is entirely understandable that those in whom a sense of the sapiential heritage of mankind is preserved are hostile to Darwinism and to so-called 'spiritual Darwinists' such as Teilhard de Chardin. But like other heresies Darwinism is a perverted truth rather than a complete falsehood. It would be helpful if this point were better appreciated in the on-going debates about evolution and religion.

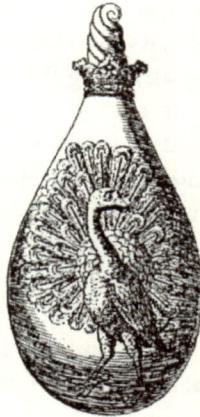

PART TWO

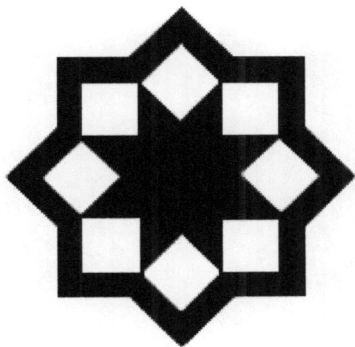

The Alchemy
of Traditional Foods

And wine that maketh glad the heart of man,
and oil to make his face shine,
and bread that strengtheneth man's heart...
Psalm 104:15

ONE of the symptoms of the modern malaise, and one of the consequences of the loss of integral wisdom, is the confusion that abounds among people regarding food, diet and nutrition. It is generally recognized that modern, industrial foods no longer nourish man as foods in the past and traditional understandings of the proper sustenance for human beings have, in all but a few parts of the world, disappeared. Modern, industrial foods, processed and prepackaged, are increasingly the norm. It was calculated several years ago that a new McDonald's restaurant opened somewhere on the Earth every seventeen hours of every day of the year, and the rate is probably faster today. Against this, there is a general hankering for more healthy foods and dietary habits, but this is a response to concerns developed out of the methods and findings of profane science, not out of an acknowledgement of a lost legacy of traditional understandings. Similarly, many people are turning to vegetarianism and more humane diets, but this is out of a squeamish horror for new 'factory' methods of meat production, not out of an appreciation of the degree to which modern meat production violates and shuns traditional notions of sacrifice and respect for the slaughtered creature. Many are dabbling with exotic diets and even more are falling into various fads. The scientists one day report that wheat grass cures cancer; the next day that it causes it. More generally, people have lost the traditional contacts with the past communicated through family recipes and food lore and the institution of the common table and, increasingly, even the knowledge and skill to prepare sustaining meals from simple ingredients. This is all a measure of alienation and of modern man's rupture from the guiding patterns of tradition.

Cooking and Cosmology

In a traditional order the preparation and understanding of food is a cosmological art and science. It is kept within its own proper dimensions and provided with a sacred context by some manner of revelation. In the Judeo-Christian world the Book of Genesis provides man with the right to enjoy the produce of the Earth—though he must labour for it—and both the Jewish and Christian religions derived from the Biblical revelations are subsequently concerned, as a matter of doctrine, with the sanctity of food. In traditional cultures everywhere the preparation and eating of food is heavily ritualized and subject to divine regulation. In the modern order, in contrast, food is simply a matter of utility, fashion and sensual indulgence. The practice of giving thanks to the Creator before a meal persists among marginalized groups of religionists but on the whole modern man gives not a thought to where his food came from and has not an inkling of its relevance to his spiritual as well as physical well-being. The Semitic avoidance of pork seems a silly superstition and the dietary strictures of Lent seem ridiculous inconveniences. More comprehensively, modern man has no idea that there was once in the past, and is still in some corners of the Earth, a science of nutrition derived from a sacred cosmology. The strictures and taboos of religions set boundaries; within those boundaries, in traditional cultures, flourish cosmological arts and sciences based upon a sacred understanding of nature. The modern health food movement is correct to point out the short-comings of modern man's divorce from nature, but his divorce from the sacred was its prelude. The health food movement is a profane reaction to the obvious inadequacies of the modern diet; it thinks in terms of chemical constituents and vitamins. In the traditional mind 'nature' is, more importantly, *Creation*—foods are evidence of God's mercy and bounty, and the natural order reflects a sacred design with an exact relation to the human being. Typically, the body of man is seen as a microcosm of the greater *kosmos*, with both permeated with an identical order that is itself of divine origin. When modern man sees a traditional Chinese meal being prepared he may think no more than 'Yum! I love stir-fry!' The health food enthusiast may take stock of the meal's protein content, minerals and enzymes and feel satisfied, in a sentimental way, that it is full of 'natural' ingredients. But a traditional man sees the bowl of the heavens in the smooth, black concave form of the wok, and he sees the grains of rice as stars and the vegetables—parsnips and carrots cut as half-moons or hexagrams—as representatives of the planets. He sees the stirring and agitation of the ingredients as mimicry of the swirling courses of the heavenly bodies and the whole act of cooking as a cosmological process in miniature. It is an act that participates in the processes of a divine and intelligent creation. Traditional approaches to foods place them within a wider cosmological context.

Categories of Food

In contrast to the approach of profane chemistry with its carbohydrates, anti-oxidants and the like, traditional approaches understand the virtues and vices of particular foods in terms of cosmological categories such as the Yin and Yang of the Chinese. Some foodstuffs are classified as Yin and some as Yang and the balance of a diet is determined by avoiding too much of one or the other or by countering foods that are strongly Yin with others that are strongly Yang. The traditional European equivalent to this was the system of four humors inherited from ancient Greek sources. Some foods were regarded as hot, others as cold, some as dry and others as moist, and their nutritional value was assessed in terms of their action upon corresponding hot, cold, dry and moist humors and organs of the body. The scientistic mind dismisses these systems as fumbling attempts to uncover the secret order of nature revealed at last by the chemists and geneticists. In fact, these systems were aspects of a profound sacred science transmitting the wisdom of an ancient contemplation of nature rooted in metaphysical principles. Something of the four humor system still exists in the Muslim world where foods are described as either hot (*garmi*) or cold (*sardi*), with four possible degrees of each, with foods acting upon either the blood, the phlegm, the bile or the black bile of the body. The classifications are not made on the basis of crude chemical analysis but refer rather to essences (*akhlat*). The heat and the cold are not measures of calories or energy with which the modern physical sciences are concerned but are cosmic polarities inherent in all things of creation. Sometimes the shape, color, habit of growth or other factors are crucial in determining the value of the food. Thus, for instance, plump, short-grained rice tends to be a hot (*garmi*) food, but the longer grained varieties are cooler. Sometimes these determinations arise directly from the recommendations of the Traditions of the Prophet Muhammad, many collections of which are particularly rich in food-lore. In one hadith, the Prophet is reported as having said, 'The main cause of disease is eating one meal on top of another.' Apart from the obvious good sense of this saying, the Muslim tradition has taken these words to refer to the mysteries of digestion which depend upon the hot and cold essences. In particular, there must be an appropriate amount of heat for the body to accomplish all of the transformations of digestion; otherwise the body grows cold and the transformations cease. Modern processed foods and the mainstays of the modern Western diet—refined sugar, refined flour, beef, dairy products, potato starch—are cold foods; typically the modern Western diet yields only enough heat (*garmi*) to accomplish the crudest physical transformations while the more subtle but vital qualities of the food remains undigested and pass through the body. Consequently, people living on such a diet seek out stimulants and turn to modern medicines which are all hot (*garmi*) in their effect. Eastern traditions such as Ayurvedic or Taoist medicine make the same analysis in terms of their own categories. The Muslim tradition finds its authority in the Islamic revelation, and uses terminology and concepts that are meaningful within Islamic civilization, but the principles are universal. Garmi and Sardi are essentially the same principles as Yin and Yang, and the Eastern traditions too understand nutrition as an alchemy of digestion requiring a balance of these universal forces.

Traditional Cooking Methods

One of the characteristics of traditional methods of cooking is the tendency to employ a small measure of heat for long periods of time. This is particularly the case with grains which were often simmered for days before being consumed. This was not only to soften the grain—a shorter period of cooking would achieve that—but also to make it more digestible and sustaining and to effect profound changes in the substance of the grain. The modern, scientific approach to these matters reports the loss of vitamins and chemical nutrients in long cooking and recommends raw or lightly cooked foods. The pace of modern life also tends to promote quick and simple meals, zipped open and popped into a microwave oven. But in China, India, Japan, the Middle East and in medieval Europe meals were often cooked for long, not short, periods and special qualities were said to have been imparted to foods prepared in these ways. Time was considered an essential factor in nutrition. This is still recognized in the case of foods like cheese and wine, which mature over time, but it is no longer recognized as important to the preparation of grain and vegetable foods. Traditional methods, found throughout the world, typically take a whole grain such as wheat berries, cover in water or broth, add a little salt, and seal in a heavy pot cooked over a very low heat overnight or for several days. Other ingredients may be added at particular stages of the cooking. Jewish cuisine knows several dishes cooked for seven days, including the proverbial Chicken Soup where a whole bird, head to feet, is boiled slowly for seven days until it is reduced to a gelatinous liquid. This is indeed a type of domestic alchemy. It recalls the long, slow cooking methods employed in the transmutations of the alchemist. This is the dimension of which modern science knows nothing. Traditional long cooking methods seek to transmute food, not just warm it through. A similar intention lies behind the Chinese practice of pickling eggs for extraordinary lengths of time, sometimes hundreds of years. The egg is not just pickled, but transformed into a new substance. These methods of food preparation are calculated to manipulate garmi or sardi, yin or yang, and to transform the essence of foodstuffs, not only their crude constituents.

Balance

The diets of so-called primitive peoples, hunter/gatherers, is characterized by diversity. Neolithic remains preserved in peatbogs reveal that people then ate a fantastic array of seeds, roots, tubers, grubs, insects, flowers and herbs, as well as fish and meat, all in small quantities. No foodstuff predominated. The advent of civilizations, however, brought the domestication of cereal grains which were used as staples. The diversity of the primitive diet became the array of accompaniments—sauces, dips, salads—to the bed of grain, rice, wheat, barley or maize, that formed the foundation of the meal. The inherent balance of the diverse primitive diet was maintained by devising methods of concentrating and enhancing foods. (The Jewish Chicken Soup, for instance, distills a foraging bird with a naturally diverse diet down to its essence.) Typically, meat consumption was irregular and connected with religious observances; a legume accompanied the grain as a staple source of protein. Rice and the soybean were the nutritional foundations of Chinese civilization, like wheat and the chick pea in the Middle East and rice and dhal or lentils in India. The modern, industrial Western diet, however, deviates from this norm significantly. Meat—devoid of all religious associations and prepared without responsibility to the creature or its Creator—has become the focus of the meal, accompanied by a narrow selection of vegetables and, very often, no grains whatsoever. The side dishes have become the main event. The 'balanced' meal described by modern nutrition experts is balanced in terms of crude chemistry but not in terms of the equilibrium crafted by the grain-based diets of the great, traditional civilizations.

The Potato

The single most disruptive historical event bringing Western diets out of step with traditional diets was the introduction of the potato from the New World. Its introduction coincided with the era of skepticism and materialism and the revolt against tradition. On a practical level, this member of the nightshade family, poisonous in every part except its tuber, became established as a *grain substitute*, and from that time forth the European diet deviated from the traditional grain-based diet. In some countries princes legislated to ban the traditional grain crops such as rye—disrupting century-old patterns of agricultural life—and to make the growing of the potato compulsory because, as a tuber, growing below the surface of the soil, it was a crop relatively immune to the destruction of invading armies. In some parts there was widespread resistance to the introduction of the potato and suspicion about its value as a food. Modern science reports on its starch and vitamin content, but the traditional mind is more concerned with the fact that, unlike the sun-loving (vertical-growing) cereals, the potato grows by division (horizontally) in the darkness of the soil and, in fact, hates the sun so much it starts producing toxins in its skin on exposure to light. Photosynthesis is a toxic process in the potato; in contrast to the grains it replaced it is a plant of the darkness.

Bread

Nothing illustrates the decline of the Western diet from traditional norms more dramatically than the recent history of bread. The white, fluffy stuff found on modern tables, alleged to be 'bread', bears no resemblance to what was known as bread in traditional times, the 'staff of life'. Modern bread is a highly processed product that, if it contains any goodness at all, has had it added in the form of synthetic vitamins and reconstitutions of the very substances destroyed during processing. The history of the decline of bread, however, has largely to do with rising agents and leavens. Traditional breads were either sourdough risen or unleavened. They were heavy and chewy and nutty in flavour and deeply sustaining. In the 19th C. German chemists isolated strains of active yeasts and the era of industrial bread, produced with a uniform rising agent, began. The sourdough was a very local foodstuff. A batter is exposed to the air and to whatever yeasts are in the region and allowed to turn sour. This is then folded into the dough and left to rise after kneading. In this way people developed a very specific acquaintance with the microflora of their district, as the scientists would explain it. Industrial or 'German' yeasts were produced in laboratories under controlled conditions; the yeasts were pure monocultures and allowed a uniform end-product that was lighter than any bread made from the comparatively haphazard sourdough method. Further refining of the flour allowed an even lighter bread until it became more of a confection than a bread in any proper sense. Very soon, the taste and texture of traditional breads was forgotten altogether. In more recent times this process has gone even further so that now chemical rising agents have replaced the German yeasts, ostensibly because they do the same job more cheaply and more quickly and because, increasingly, people are intolerant to the industrial strains of yeast. The sacred status of and the mysteries associated with bread in various traditions need not be recounted in detail here. We need only point out what an extraordinary spectacle it is that the decline of bread sketched above could have occurred in a civilization that in ancient times knew the cult of Demeter and the mysteries of Eleusis and, for the last 2,000 years, has had the Christian Eucharist, a meal of bread, as its central and most solemn ritual.

Drinks

A number of other things also deserve to be mentioned as symptomatic of the way dietary changes in the modern West parallel its deviation from tradition in general. Modernity is thirsty; both metaphorically, for a wisdom it no longer even suspects exists, and literally for a diet awash with drinks. Other than the proliferation of McDonald's restaurants, the global reach of Coca Cola promotion is another emblem of the spread of non-traditional, industrial foods. Traditional diets include far fewer liquids than the typical modern, Western diet. This is largely because traditional diets are grain based and a large portion of the necessary daily liquids is consumed through the liquid absorbed through boiled or steamed grains. There are far fewer liquids and far more salts in meat, however, so the modern meat-based diet requires supplementary liquids, usually in the form of sugared drinks and soda waters or plentiful cups of tea and coffee. In traditional cultures, such as that of Japan, drinks come in small cups and are infrequent. The notion that drinking large quantities of liquid is a healthy practice is non-traditional. Traditional understandings think of the human digestive system, as indicated above, as like the *athenor* of the alchemist; popular opinion in the modern West holds it to be a type of drainage system that needs to be flushed out regularly. The subtle transmutations achieved by traditional cooking methods are designed to duplicate and advance the processes of human digestion. (In many languages the same word is used for cooking and to describe the processes of digestion.) Many traditional foods, such as soybean foods like the Japanese *miso*, or dairy foods such as yoghurt, or brassica foods such as sauerkraut, are predigested ferments specially adapted by traditional methods for human digestion. Their benefits, and all but the crudest processes of digestion, are lost in a diet with a high liquid intake. Alcoholic beverages, wine and beer, were once foods, means of preserving juice and grains, again by means of live yeasts and fermentation. Their decline and denaturing is evident from the fact that these drinks today require artificial preservatives to keep them, a task originally and properly belonging to the alcohol itself.

Salt

Related to the high liquid intake of the modern diet is a profoundly disturbed relationship between man and his most intimate contact with the mineral realm, salt. Salt is the traditional, universal condiment of mankind, essential for his survival and for his enjoyment of flavour, yet in modern nutritional reckonings it is problematical and associated with various modern diseases. This has to do, again, with the refining of the modern table salt and its corruption with free-flowing agents and, again, preservatives (as if salt, like alcohol, were not a preservative itself), and to do with the high consumption of meat and animal-grade salts in the modern diet. Traditional cultures understand better the proper, even sacred, role of salt in the life of man and use it appropriately as a catalyst to enhance the flavour of food rather than to mask its tastelessness, which is the role of salt in industrial, processed foods. Salt is a bond of human community. When Semitic food restrictions forbid the drinking of blood they are, just as much, insisting upon a social order that ensures the proper distribution of mineral or sea salt. According to the *hadith*, the Prophet Muhammad began and finished every meal with a pinch of salt and praised it as a blessed thing. It is a symbol of purity and wisdom. Without the catalyst of salt the transmutations of traditional cooking methods are without effect.

Conclusion

Interest in Eastern spirituality among dissatisfied Westerners is often accompanied by the discovery of a different and more traditional type of cuisine. This is a very immediate way in which they can experience something of the traditional order for which they long. The traditional culinary wisdom of the Chinese and Japanese no longer prevails in China and Japan, but it is far more intact than traditional ways in the West. With the necessary adjustments it is still possible to reconstruct many traditional methods and recipes from their current corruptions. This is also true of Middle Eastern cuisine. It has been corrupted with cane sugar (a cold—*sardi*—form of sweetness, sweeteners normally being hot foods), with stimulants (coffee), nightshades (tomatoes, eggplants), and, as in Islamic culture generally, affluent urban living has led to the over-consumption of meat, but it is still possible to discern the outlines of the traditional diet, based on cereal grains (cous cous, burghul) and legumes (fava beans, chick peas). In rural areas it is still possible to find people cooking rich, grain-based stews using long, slow traditional cooking methods. These are the same areas where traditional craftsmen can still be found crafting their wares and where sacred patterns, derived from revelation, inform every aspect of life. They will tell you in these parts that the fast of Ramadan is not only to remind the faithful of what it is like to be hungry—a

sociological and sentimental explanation—but that it has mysterious effects upon the liver and the humors, and they will rec-
ommend traditional Ramadan dishes that cleanse the organs of
the body and bring visions to the soul. There can be no doubt
that modernity brings with it a diet that is not only a product of
profane understandings but that makes men profane beings,
insensitive to the spiritual and isolated from the living forces of
the *kosmos*. There is more at stake in the foods we eat than refill-
ing our inner test-tubes on a regular basis or of avoiding carcin-
ogens and other hazards, and there is more to food in a sacred
culture than simply saying 'Grace' with meals. Industrial foods
are fodder for automatons, soulless food for the soulless. This is
not just a matter of health, but of a relation to the macrocosm,
our place as creatures in creation.

The Man-Plant

Central Themes
in the Alchemy of Farming

As regards the nobility of birth... the forefathers of these men were not of immigrant stock, nor were their sons declared by their origins to be strangers in the land sprung from immigrants, but natives sprung from the living soil... and [were] nurtured by that mother-country wherein they dwelt, which bore them and raised them and at their death receives them again to rest in their own abode...
During the period in which the whole earth was putting forth and producing animals of every kind, wild and tame, our country showed herself void of wild animals, but chose for herself to give birth to man...
Now our land, which is also our mother... of all lands which then existed, she was the first and only one to produce human nourishment, namely the grain of wheat and barley, whereby the race of mankind is most richly and well nourished, for she was herself the true mother of this creature.
                                                        Plato, *Menexenus*, 237–238

WHILE it is well known that the craft of smithing and the arts of the forge were, from very ancient times, accompanied by an esoteric science that has come down to us as that body of knowledge called alchemy, it is less well appreciated that other crafts, and especially the craft of farming, were accompanied by parallel sciences, all expressions of a traditional cosmology. Alchemy, since it specifically concerns metallurgy, continues to hold some curiosity for industrial man, but the very notion that farming once transmitted, along with knowledge of skill and technique, an esoteric, spiritual understanding of the farmer's craft, is all but lost in an urbanized world. This is despite it being obvious that the world's great traditional mythologies are agriculturally based, as indeed were the ancient mysteries such as those at Eleusis which enacted the agricultural myth of Demeter and Persephone. The importance of the smith to these civilizations is acknowledged, but it is little appreciated that the primary craft, and the craft the work of the smith served, is farming, and that it occupies a central place in traditional cosmological understandings with an esoteric symbolism of its own. The purpose of this short article is to sketch the main themes of this symbolism, especially as they are interconnected with and compliment the foundation mythology of the alchemical tradition. The example to which we shall refer is the agriculture of ancient Attica, along with the mythology and rites of the Acropolis which, it should always be remembered, extend back well beyond historical times.

There is an ancient, primordial mythology found throughout many parts of the world, and rehearsed in Hesiod, concerning the original 'Golden Race' of human beings who enjoyed a distant 'Golden Age'. In this mythology time is cyclic and the Golden Age eventually gives way to a Silver Age, then a Bronze Age, and finally an Iron Age which prepares the way for a new Golden Age and a new cycle. In many versions of this mythology the change from one cycle to another involves reversals such as the Earth changing its direction of rotation. In many mythologies, including those Biblically based, at the end of the Iron Age the dead will be resurrected from their graves. More commonly, in a parallel idea, the human beings of the Golden Age are described as being born from the soil *like plants*. Often, again, the resurrection motif appears in the serpentine form of the plant-man. This myth was the basis of the autochthony cultus of ancient Attica. The Atticans boasted that, unlike other Greeks, their distant ancestors were not immigrants from other lands but had grown—like plants—from the native soil, autochthons, aboriginal in the fullest sense. At birth the children of Attica were given a gold, serpentine necklace as an emblem of their autochthony, and similarly the golden arm and head-bands characteristic of the Athenians alluded to their autochthony, their birth from the Attic soil, the basis of their nobility. An ancient, spiritual understanding of agriculture and agricultural practices accompanied this mythology. It was both a theory and a praxis, the secret craft of the farmer, the farmer's alchemy. In the Attic version of the great myth of the nuptial of Sky-god and Earth-mother, their father is Hephaestus, the Olympian smith god, and their mother is Gaia, fertile Earth. The rites of the Acropolis featured sacred ploughings to celebrate their union. The smiths and metalworkers celebrated the cult of Hephaestus. The farmers of the Attic soil celebrated the cult of Gaia.

The objective of their agriculture, as we must infer from their rituals and myths, not only concerned quantitative production—pleasing the gods to give forth their bounty—but the inner quality or cosmological status of the plants grown in their black and sacred soil.[1] This, the Attic farmer believed, was the very soil from which the primordial Golden Race had sprung like plants, and so we can imagine what the same Attic farmer, employed in his daily craft, must also have believed of the plants growing under his cultivation. In this view, ploughing and tilling the soil are cosmological acts and the craft of working the soil is a cosmological art referring at every turn to higher realities. When the farmer's ploughshare penetrates the fertile soil, exposing the Earth to the Sky, he reenacts a cosmogony. And, by this particular mythology, he seeks to grow plants of an *aurumic* quality,

---

1. The soils of Attica are today wasted and their decline was already advanced in classical times, and it may be objected that even before that they were never particularly rich and fertile, except in contrast to surrounding districts. Yet from these same soils came the cult of Demeter and a comprehensively agricultural mythology and religion. The Classical Greeks report that the Attic soil was one—long ago—much richer and so, in a sense, they celebrate a legendary fertility. But beyond this it needs to be noted that the Greek doctrines outlined in this article were, most likely, transplanted to Attica from the Nile delta. Greek sources mention the region around Sais in particular.

golden in their essence. This was another boast of the Atticans—the quality and taste of their produce, their golden grain and their golden honey, was unsurpassed in all Greece. The full symbolic importance of this aurumic quality, however, extends to a spiritual and not just sensory excellence. The aurumic plant is not just fine produce: it retains the quality of primordiality. So, in practice, the aurumic plant is the plant that is *man-like* or has a certain *human* quality. This, strange as it may sound to modern ears, is the great secret of the traditional farmer: he seeks the *man-plant.* Just as the 'Golden Race' had been born from the soil like plants as plant-like men, so in this Iron Age the farmer seeks the inverse—man-like plants. Farming, then, is a type of inverted autochthony. The aurumic plant is the plant that grows as much like a human being as possible. Modern readers will probably know something of the folklore surrounding ginseng in Asia and the mandrake root in Europe; such folklore reflects something of this tradition, but the full order of cosmological symbolism that is brought into play here has been lost in modern times.

It is necessary to understand, for a start, that, in this view, plant and man are exactly inverse: one the macrocosmic creation and the other the microcosmic paradigm. Man is both upside down and inside out in relation to the plant. In Plato's cosmology—which itself is based on the Attic mythology sketched above[2]—we are given the image of the human body as an upside down tree.[3] The head of man is parallel to the root of the tree, the gut and genitalia of man parallel to the flowers and fruits. In their organizational structures, plant and man are polar opposites.[4] The good plant, then, is most man-like not when it most *resembles* a man necessarily but when it is most perfectly

2. As a careful reading of the dialogues *Timaeus* and *Critias* will reveal. The Platonic Demiurge is clearly based on the mythological Hephaestus (of the Attic cult) and the 'Receptacle of Becoming'—called 'mother and nurse'—on Gaia. The Athena cultus forms the background of Plato's doctrine of the World-Soul.

3. See *Timaeus* 90a.

4. 'God has given each of us, as his daemon, that kind of soul which is housed in the top of our body and which raises us—seeing that we are not an

inverse—an inverse reflection—to man, in its organization. So, paradoxically, the man-plant is always the plant that is most itself, most nearly expressing its own essence. Farming enacts the nuptial of Earth and Sky. The aurumic plant is that plant that stands, as it were, newly created on the very day of Creation, both an anticipation and a recollection of the Golden Race.

In the praxis of farming this aurumic perfection is sought through traditional methods of what would today be called 'organic' cultivation and soil husbandry. And it manifests most obviously in the presence of a certain 'intelligence' in plants grown in soils worked by these methods over long periods of time. Modern, profane science explains that so-called 'organic' produce tastes and keeps better and has other superior qualities because plants grown in organic soils are 'free to choose' their nutrients rather than being force-fed by water soluble chemical nutrients as is the case in industrial agriculture. The truly 'organic' plant is nurtured on a stable soil humus which allows it to take nutrients as required in reaction to its environment, and specifically in relation to sunlight. This is a wisdom denied to the force-fed industrial plant. Industrial plants are equivalent to senseless automatons. The organic plant, by contrast, seems to 'think'.[5] This is the quality the traditional farmer seeks. Grown in the right soils the aurumic plant seems possessed of a presence and an intelligence that is nowhere in itself yet seems profoundly akin to our own, plants that seem strangely human. The relation between the word 'humus' and the word 'human'[6] still

earthly but a heavenly plant—up from the earth towards our kindred in the heavens.'—Plato, *Timaeus* 90a.

5. Parallels to this and other themes covered in this article will be found in Rudolf Steiner's lectures on Agriculture and embodied in Steiner's 'biodynamic' system of agriculture. This is a specialized 'organic' system with what might be called 'alchemical' features. While biodynamic produce goes by the international trading label 'Demeter', Steiner makes no mention of Greek sources for his ideas in his lectures and was probably drawing on quite different (northern European) sources which illustrates the universality of these themes.

6. From the common Indo-European root *ghom* meaning soil.

signals this idea in modern languages. Plant life, of course, possesses no inner life separate to its environment. Plants have no
internal organs. Their organs are the Sun and Moon and Stars. In
this they are the macrocosmic counterpart to man, the completed microcosm. Plants cannot 'think' as men do, yet the aurumic plant, when it is fully inverse to man, possesses the *nous* of
the World-Soul. This is the plant that stands in balanced relation
between Sky and Earth. Traditional 'organic' farming was not
only concerned with maintaining an ecological stewardship and
such profane concerns as sustainable levels of production; traditional farming was part of a total cosmology and the farmer's
specific task was to grow plants with man-like intelligence. This
is all to be understood in relation to the myth of the Golden Age.

When this general background is understood, other symbolisms become clear. It is pertinent to consider here, for instance,
that 'gold' throughout this mythology is in fact a symbol of and
metaphor for *light*. What we have called the aurumic plant
might otherwise be said to be cosmologically *translucid*. In
alchemy gold is a condensation of light, the solar essence in
embryonic form. The alchemist seeks to extract this embryo or
seed from its womb, the Earth, and hasten its maturation. Thus
does the alchemist do the Great Work, advancing the Golden
Age. In farming, the farmer works with four condensations of
primal light: sunlight, air, water, and crystal, the silica of the bedrock in whose crystalline womb seeds of gold ore are so often
found. The traditional farmer in ancient Attica and elsewhere is
not thinking in terms of nitrogen and potassium or of microbes
and Ph, but in elemental terms. The act of ploughing, while
sharing an obvious sexual symbolism with alchemy, is also conceived in terms of bringing the higher and more rarefied elements to the lower and denser ones. Specifically, ploughing
brings the sacred, aurumic element *light* into the soil, and the
farmer, demiurgically, sows his soil with an Apollonian essence.
This manifests in what the agricultural scientist would call
'improved soil structure'—the patterns of the Sun textured in
the root-paths of plants. Improved aeration and water-penetra-
tion also follows tillage: here we must realize that air is to be

understood as a condensation of light—like light it is invisible. And water, in turn, is a condensation of air and shares with light and air the quality of transparency. In any case, the farmer brings the celestial element light—and its condensations—into the Earth, thereby fertilizing her. Akin to this is the traditional practice of tilling for frost. The farmer lightly tills the soil before the frost is expected. A heavy frost on freshly tilled soil will improve structure and 'tilth' by freezing and thus expanding the soil water and causing the soil to gently crack along its lines of least resistance. Here the traditional farmer is working with the same symbolism that makes frost and dew important—as stellar condensations—in the alchemical tradition. The farmer acts as/ with Sky-god: he brings the Sky to the Earth, the cosmic order to the passive, maternal soil. Moreover, the black soil and its teaming life is a reflection of the night sky and its teaming stars and the cycles of the land are analogous to the cycles of the heavens, and all cycles are ultimately subsumed in the cycle of the Great Year from one Golden Age to the next.

A further symbolism that is crucial here is that, in an important sense, namely the 'point of view' of the living plant, the Earth herself shares the same organization of the plant, and so her 'head' (so to speak) is below the surface and her 'stomach' and 'womb' above. This is inverse but complimentary to the symbolism of metallurgic alchemy. More commonly we say the womb of the Earth is below the surface, but actually this is the 'point of view' of the embryonic metals and is not immediately relevant to the farmer who is nurturing plants. In the farmer's perspective, the head or root of the living Earth is below the surface, in the bedrock. The stomach and womb of the Earth, then, where the Earth flowers and fruits, is the atmosphere above the surface. In this sense farming takes place in the 'belly' of the system. Moreover, the celestial archetypes are not mediated to plants directly from the sky, but through the 'head' of the Earth: that is, they are reflected upwards from within the Earth. Chthonic and celestial forces are, finally, interchangeable and they have been interchanged in this instance. Thus the traditional farmer, by cultivation, assists in the deep rooting of plants that they might have a strong connection to the Earth, and he cares for the soil in which the plant grows, in order that they may manifest more truly their *celestial* archetypes. The traditional farmer knows, intuitively if not consciously, that a living soil expresses the celestial forms and that a plant rooted to the Earth is also, in the same measure, rooted to heaven. In Attic mythology this interchangeable symbolism introduces Athena as a third party between Hephaestus and Gaia along with the story of her surrogate motherhood of the golden, autochthonous child, a motif that the people of Athens made their own. In this formulation Athena is essentially an Air and Cloud, or atmosphere, deity.[7] The famous depiction of her receiving the solar child from Gaia, to act as its foster-mother, by the Kodros Painter, relates the interchangeability of the terrestrial 'womb' which may either be conceived as being below or above the surface.[8] The primordial plant-man has two mothers. Moreover, the same interchangeable symbolism is found in the story of Athena's own birth not from a womb but from the head of Zeus. The practical operations of farming actually depend upon this symbolism; the farmer has a primary concern in the care of the roots of plants, for they are the plant's connection to the *heavens*. The traditional crops of Attica—vine, olive, fig—are all deep-rooted sub-soil exploiters: thus do their fruits contain oils and essences of a sacred character, extracted from the Sky through the Earth.[9]

The role of Athena in the Attic mythology deserves further

7. This is important but often neglected. The fineness of the Athenian intellect was attributed to the peculiar clarity and quality of the Attic air. There is an Air symbolism—parallel to soil symbolism—throughout Athena's mythology. One of the best accounts of this is a little treatise by the English artist John Ruskin entitled 'Athena, Queen of the Air'.

8. Although in the iconography of this scene recorded in Athenian vase painting it will be noticed that the 'surface' is depicted by two spears crossed at the angle of the solar tropics, alluding to a yet more esoteric set of parallels having to do not only with the cycle of the seasons but also the cycle of rites in the sacred calendar of Attica.

9. The ancient Athenians had as their emblem and cultivated in large and prolific nurseries outside of Athens the violet (viola odorata). In this context the plant actually has the same symbolism of gold, only by virtue of violet being

comment here because in many traditions the Sky/Earth duality is seemingly complicated by a third principle. In fact, the character and function of this third principle is simple in itself, but it introduces the potential for complex symbolic development, which is certainly what we find in the stratified archaeology of the Acropolis.[10] To reduce it to simple terms: the third principle is astronomically speaking the Moon (between Sun and Earth), the cycles of which are of intimate practical concern to the farmer. Residual forms of 'moon-planting' can be found throughout the world, and even in the affluent West, but the inner significance of this practice has been all but lost. In ancient Athens its fullest development was in the doctrine of *seasonable marriage* as it applied to citizens of the state, of which doctrine Athena was patroness. The key to autochthony, this doctrine held, was conception and birth at cosmically propitious times. The marriage feasts of the Athenians were calculated so that children might be born at New Year, the birthday of the goddess, analogous to the day of Creation, in this way perpetuating the aurumic quality of citizens. This astrological eugenics, the decline of which Plato describes as the cause of the decline of the Ideal State,[11] declares that time of conception

opposite to that color (for which reason too purple has regal and noble associations and an inherently aurumic symbolism). The violet viola tricolor has a significant but little appreciated alchemical symbolism concerning the sporting of three colors in its flowers, synthesized into violet.

10. We are not concerned here to delineate or illuminate the vast tangle of mythological and cultic motifs in ancient Greek religion but only to point out a central theme which occurs and reoccurs in both explicit and subtle ways in our literary and archaeological sources.

11. In the so-called 'Nuptial Number' of the *Republic*, Plato's most famous mathematical conundrum. The cause of the decline of the Ideal State is that its citizens begin to be born at irregular times which causes a decline in the quality of their souls.

and birth, not bloodline, is the mechanism of the nobility of autochthony. Similarly, the farmer—who seeks the man-plant—sows his seeds not only according to time of year and such local factors as the frost-free date, or according to breeding lines, but according to the configurations of the starry heavens, because the spiritual, aurumic quality of plants is attuned thereto. In the mythology of the Golden Age, conceptions and births were once at appointed times. But in the skewed conditions of the Iron Age 'seasonable marriage' must be calculated. Just as 'seasonable marriage' confers the quality of autochthony on citizens of the state, so knowledge of *when* to sow is critical in the farmer's quest for the aurumic plant.

In the sub-lunary realm the third principle, represented by Athena in this mythology, is the airy and moist elements, the atmosphere, clouds and weather, that separate the celestial order from the terrestrial (just as the Moon mediates between Sun and Earth). In the Attic surrogacy myth, Hephaestus ejaculates on Athena's thigh—without violating her virginity—and she wipes it off with wool which she throws to the Earth, thus impregnating Gaia. In this mythologem, quite simply, the nuptial of Sky and Earth is expressed as the Sky God depositing his seed in the clouds (wool) from where they are distributed over the Earth as fertilizing rain, guided by Athena's nurture and wisdom. The goddess' mighty shield—one of her principle attributes, denoting defensive warfare—is emblematic of the disk of the full Moon but is, at the same time, within the chain of associations operating here, nothing less than the atmosphere that shields the Earth from the potent rays of the Sun—the ozone layer and so forth, as it is described today. The motif of Athena's productive virginity—she is conspicuously Athena Nike, triumphant in her chastity—suggests an elemental ecology: namely that as the farmer tills the soil and brings the solar, celestial element to the Earth, the intermediate elements must facilitate but emerge unscathed by the exchange. In profane terms, agriculture must proceed without befouling air and water, the vehicles of life. The virgin motherhood of Athena is different to the all-yielding motherhood of Gaia and celebrates a different ideal: it explicitly underlines the idea that the farmer extracts bounty from nature without diminishing or defiling her. In traditional agricultural practice this has another significance too, one that might come as a surprise to the modern agronomist; the traditional farmer draws fertility *from the air*, not directly from the earth. He does this most obviously through including legumes in his planting cycles, which draw nitrogen from the air and fix it into the soil. A traditional farm unit can maintain production and export produce without having to import fertility and can continue doing this indefinitely—an image of productive virginity. This defies the thinking of industrial agriculture which considers the farm as a type of test-tube *in situ*, but the traditional farm is an open system drawing indefinitely on the nitrogen and carbon of the atmosphere—not 'stripping' or 'mining' the soil—as the source of nature's wealth. One of the unforeseen consequences of industrial agriculture has been profound disturbances in the atmospheric carbon of the Earth, with consequent changes in climate and weather. In mythological terms, it is an Hephaestean agriculture that has violated Athena; it is an agriculture of force, not of wisdom and skill. It remains to be said that the wisdom and skill of Athena has a more esoteric significance than simply judicious and sustainable agriculture. Beyond these things she is goddess of the World-Soul and its harmonies, and in truth hers is the *nous* with which the aurumic man-plant 'thinks'. It is of her wisdom—a macrocosmic intelligence—that the aurumic plant partakes.

The aurumic plant also has a number of other features which are signaled in Greek mythology. One of these is that plants grown in ideal 'organic' conditions are more robust in terms of their resilience against macrocosmic intrusions. Specifically—as is well known today—such plants have greater resistance to disease and pest infestation, but also against extremes and variations in soil conditions as well as weather and climate. For example, 'organic' plants—drawing upon a rich store of soil humus—can tolerate wide variations in soil acidity or alkalinity (Ph) and such plants have greater resistance to frost and cold and to extremes of heat. The aurumic plant, in short, is adaptable.

This is a sign of its intelligence. Most importantly, in practical life and for the purposes of the cosmological symbolism we are pursuing here, it is possible to actually *extend the growing season* by growing stronger, healthier plants. The traditional farmer did not fail to notice this: the more aurumic the plant, the less it is bound by the seasons. This reminds us of one of the characteristics of the Golden Age: it was seasonless. Homer presents a beautiful image of the aurumic garden in his description of Phaeacea:

> And without the courtyard hard by the door is a great garden, off our ploughgates, and a hedge runs round on either side. And there grow tall trees blossoming, pear-trees and pomegranates, and apple-trees with bright fruit, and sweet figs, and olives in their bloom. The fruit of these trees never perishes nor fails, winter nor summer, enduring through all the year.

The Iron Age farmer is remote from this mythic ideal, but can nevertheless labour to create the best possible conditions for his aurumic plants which, as a dim reflection of the seasonless Golden Age, will outlast inferior crops at the start and finish of a season, if only by a few crucial days or weeks.

Another detail in Homer's description of the garden of Phaeacea deserves mentioning here too, for it provides further insights into the practical applications of these themes and the full extent of the interpenetration of mythology and farming. As well as trees in the Phaeacean garden, Homer tells us, there 'are all manner of garden beds' and just as the trees are perpetually in fruit so are these garden beds 'perpetually fresh'. This alludes to the Greek horticultural practice of maintaining freshly deep-dug permanent, raised beds for the growing of flowers and vegetables. The modern scholar will explain to us that the Greeks, living as they did in a volcanically active earthquake zone, had observed that plants naturally thrive in soils that have been freshly moved by landslides. They imitated these conditions in their garden beds by maintaining a loose soil, freshly moved by cultivation. The mythological background to this practice in Attica concerns Hephaestus as the volcano god; the landslide is one of his modes of cultivation. The gardener, by *mimesis*, creates the loose, fertile conditions of a landslide in his garden beds which, moreover, are shaped permanently and are raised, resembling graves, in a further allusion to this same Golden Age symbolism.

The serpent motif that often accompanies this symbolism and is pronounced in the case of the Attic autochthony cult is emblematic of the *cycle* itself and specifically of regeneration: the snake hibernates the winter and in the spring is reborn perpetually, shedding the skin of the old year and by analogy the old cycle. When the snake motif appears in combination with the plant-man it simply signals that this is a cyclic and not a unique creation. The many extensions of the serpent motif into alchemical symbolism are well known. One of the most central and most developed is that of the fiery dragon which—in another example of the interchangeability of chthonic and celestial symbolism—guards the embryonic gold in its subterranean lair, as well as having wings and flying, devouring the Sun as the dragon of the eclipse, a sun-eater. The extension of the serpent motif into the alchemy of farming is, rather obviously, in the role of the *earthworm* in the recycling of organic wastes and in the production and maintenance of humus which is, in truth, the special vehicle of the aurumic essence. Many languages make no distinction between a dragon and a worm (or wirm or vorm, etc.) and in the alchemical cosmology the dragons that dwell in the depths of the Earth protecting their store of gold and the worms of the top soil nurturing the store of humus have an obvious relation. To the traditional observer the notable characteristic of the worm is that it appears to manifest spontaneously from nowhere; it is as if they are an implicit possibility of the soil, and when measures are taken to improve the soil they appear in plenty, spontaneously. Their presence, though, indicates, again, cycle and regeneration. Organic cultivation depends upon hastening and advancing the cycles of growth and decay—in composting, for instance—which imitate, in miniature, the great cycles of nature and the *kosmos*, and in this the worms that are encouraged by and multiply under organic cultivation are the serpentine emblems of cyclic regeneration. Life and death, as

Socrates said, follow from one another like night and day.[12] The alchemist either tames or slays the dragon in order to secure the gold. The farmer too either tames the worm—as a worker in labouring for the conditions that will yield the aurumic plant, for worms help accomplish all the tasks required—but must also slay it (cyclically) as an inevitable consequence of cultivation. The symbolism of sacred ploughing will often be associated with dragon-slaying motifs or other serpent motifs such as in the custom of ritually 'beating the bounds' of lands to chase away snakes, for example. Cultivation inevitably brings momentary catastrophe to the microcosmic worlds of the soil flora and fauna. But the soil and its life regenerates. 'The cut worm forgives the plough' as William Blake said. Ploughing, traditional farmers well knew, is, in a very real sense, always an act of violence upon the Earth. The Sky-God is often said to force himself upon the Earth Mother. The primordial nuptial is often depicted as a rape. In Attic mythology, the advances of Hephaestus upon Athena and the impregnation of Gaia were certainly uninvited, although here the violence has been made somewhat muted and reduced, we might say, to a charge of indecent assault more than outright rape. By sanctification, ritual and skill—by imitation of nature in her mode of operation, as Thomas Aquinas said—the violence of cultivation can be mitigated and the result made felicitous and holy. Ploughing cuts the worm but the worm regenerates and in fact multiplies. And not just the worms the eye can see, but also the entire wriggling universe of serpentine microscopic forms that are born and die in truly astronomical numbers and whose 'sheaths'—shed like snake-skins—actually contribute the essential cellular substance that builds a fertile, healthy soil. This soil life is built up only to be destroyed again in the necessary cycles of cultivation. Analogous to a cosmic

12. As a side note, the Pythagorean arguments for the immortality of the soul provided by Socrates in Plato's *Phaedo* need to be understood as applying to the cycles of the primordial plant-man. Note, for instance, that Socrates' argument for a life-death cycle is strangely vegetative and—as astute undergraduates sometimes observe—would 'work better' for plants than for human beings. This is true, and significant.

'ploughing', the Iron Age is characterized by destruction, with the foundations of the cosmic order torn apart and over-turned—and here the serpent symbolism often takes monstrous and menacing forms—but the new Golden Age will nevertheless blossom from the seed-bed of ruin. The serpent in its many guises symbolizes the inherent regenerative capacity of the cycle itself, both in its macrocosmic totality—where, in alchemical symbolism, we usually find the serpent swallowing its own tail—and in the smaller cycles of nature which enact the greater cycles in miniature. In practical terms, the farmer knows that the worm is a sure sign of good soil husbandry; it indicates that the 'dragon of the cycle' is present to tend the aurumic essence.

In conclusion then, the alchemy of farming consists in a pri-mal participation in the cosmogonic act, the marriage of Heaven and Earth. Essentially, the farmer brings the celestial element, light—and its condensations—to the maternal soil and thus acts as Demiurge at the work of Creation. The alchemical cosmog-ony often takes the form of the lightning strike, the ejaculatory impregnation of the Earth by the Sky-god.[13] To the farmer it is the impregnation of the Earth by light as Sun and Rain and Wind. In alchemy, the alchemist mines the Earth for the seeds of celestial gold. In farming, the farmer is the sower of the solar seeds, and his alchemy consists in bringing forth, by wise cultiva-tion, the 'man-plant', reflection of the 'plant-man' of the Golden Age. These are plants endowed with the *nous* of the World-Soul, displaying wisdom and intelligence like men, and in this are per-fect manifestations of their archetypal forms. It will be found that this mythology underpins nearly all aspects of the agricul-tural cultus of the Acropolis, especially in the layers under Egyp-tian influence,[14] as well as taking similar forms in other cultures at other times. Variations on this mythology underpin the Near Eastern custom of saving the best sheaf of wheat from a harvest,

13. In fact, lightning strikes are of benefit to the farmer too; they fix atmo-spheric nitrogen to the soil, a phenomenon that is imitated—using vast amounts of energy—to create artificial nitrogenous fertilizers.

14. See footnote 1, p105.

nominating him the 'Old Man' and burying him in a tomb with prayers, to give but one example. Elsewhere in Greek mythology the myth of the garden of the Hesperides alludes to the same pattern of ideas. In this primordial garden, the Greek equivalent to the Biblical Eden, grow golden apples, gifts of Gaia to Hera on the occasion of her wedding to Zeus. These are the golden fruits Athena had Herakles steal for her as one of his twelve labours. Here again we find the aurumic plant connected with the primordial nuptial. The motif of Athena having the fruit stolen concerns the decline of the Ages and her role—presented as a subversion of Hera's primordial rulership of marriage—in *seasonable marriage*, the twelve labours having a zodiacal signification. In modern times it is difficult to appreciate how these mythological themes translated into the actual practical life of a traditional agricultural society. The stark truth is that we no longer understand either mythology *or* agriculture. We forget that, even in Plato's time, there was farmland within the center of Athens, and the vast majority of Greeks were engaged in farming. What did these myths, and the world-view they embody, mean to ancient people of the soil? In practical terms an aurumic agriculture nurtures the soil humus and cultivates to bring light and air into the Earth, and the farmer calculates his sowings that the marriage of Heaven and Earth might be 'seasonable', and rotates legumes and employs other practices so that abundance might be taken from the air, not the earth, and though each year brings a harvest, nothing is spoilt or diminished. Most importantly, the traditional farmer did not fail to notice that plants grown in humus-rich, well-structured soils that allow for deeprooting not only grow well in terms of quantity but display a type of 'intelligence' that consists of an attunement to the macrocosm and that is equivalent to health, vitality, color and flavour. The symbolism that equates this with the 'man-plant', it will be found, underpins traditional farming everywhere, regardless of how dimly remembered and overgrown with superstition is the original doctrine. This is a side of alchemy that needs to be rediscovered. Amongst other things it important that this other side of alchemy be appreciated anew so that a

traditional critique of the violations of industrial agriculture—
which goes about farming as if it were *mining*—can be conducted within a proper framework.

Capitalism, Tradition
and Traditionalism

The weight of gold coming to Solomon in one year
was six hundred and sixty-six talents...
1 Kings 10:14

IT is remarkable that in Traditionalist writing, which on all
other accounts gives a radical and comprehensive critique of
modernity, there is lacking any systematic account of the role of
capitalism in the 'Reign of Quantity'.[1] This is remarkable
because, to the present writer, and to many others, it is self-evident that capitalism is, in our time, the principle agent of anti-
traditional action and of the reduction of the world to mere
quantities. We find some sharp and insightful criticism of
socialism and its project for a horizontal, atheistic social order
among the Traditionalists, but no corresponding critique of
capitalism, even though socialism is, certainly as Marx presented it, a sub-set of capitalism. Or, at least, socialism and capitalism are simply two sides to the one coin, so it is surprising to

---

1. The only exception seems to be the article 'Traditional Economics and
Liberation Theology' by Dr. Rama P. Coomaraswamy, appearing in *In Quest of
the Sacred: The Modern World in Light of Tradition*, ed. Seyyed Hossein Nasr and
Katherine O'Brien, Foundation for Traditional Studies, 1994.

find only a critique of one side of it among these writers. Some of the more dirt-under-the-nails Traditionalist writers, or those who themselves have a craft background, such as Wendell Berry, Gai Eaton, and Eric Gill, have tackled matters of practical economy, but otherwise, in various lists of Traditionalist demonologies 'capitalism' is conspicuously absent. We find: scientism, psychologism, evolutionism, progressivism, individualism, humanism and so on, but not capitalism. We find a thoroughgoing critique of Darwin but not of Adam Smith. At least some of the -isms in the Traditionalist demonology are, in the present writers opinion, best understood as sub-sets of capitalism too. Modern capitalism conspicuously promotes scientism, psychologism, evolutionism, progressivism, and the others and in some cases it is arguable that capitalism gave birth to these things in their modern form and continues to be their matrix, so why do the Traditionalists not turn their critical gaze to capital? We need not indulge in economic determinism to see the importance of capitalism in creating and maintaining the modern order. We can agree with the Traditionalists and see the modern malady as essentially spiritual and capitalism as merely a symptom, not a cause, but there is a hierarchy among symptoms and certain constellations of symptoms constitute if not a disease then at least a syndrome that should be identified. It might be argued that between them the major Traditionalist writers cover the whole content of capitalism even if they rarely talk about the thing itself. In that case, the purpose of the present article is to demonstrate that it is useful, from a Traditionalist position, to identify capitalism as a single, distinct monster and one whose world-devouring, anti-traditional character is far clearer since the demise of authoritarian socialism in the late twentieth century. One hesitates to call the Traditionalists 'writers of their time' but as the decades roll by and the New World Order takes shape, their concentration on the evils of socialism—evils which were certainly real—seems disproportionate and one-sided.

By capitalism, it needs to be said, we do not mean simply 'free enterprise' or the 'free market' or the 'free exchange of goods' as if capitalism is a sort of big version of the type of exchanges one finds in local marketplaces; rather capitalism is a distinct, modern ideology that is profoundly hostile to traditional, local patterns of exchange,[2] and all other essential aspects of tradition as well. The term 'capitalism' was first coined in the European Enlightenment in recognition that a certain world-view had matured into an identifiable ideology. If one consults almost any short history and definition of capitalism, one first finds that it is a product of the Enlightenment. The present writer consulted the Internet (that modern parody of the *Akasha*) on this question and arrived first at an article by a Richard Hooker[3] who explains that

> capitalism is more than just a body of social practices easily applied across geographical and historical distances, it is also a 'way of thinking', and as a way of thinking does not necessarily apply to earlier European origins of capitalism or to capitalism as practiced in other cultures.

While we might find *capitalistic* elements in other cultures and other periods, capital*ism* as an ideology, as a way of thinking and as a world-view is a solidification of the European Enlightenment and hence part of the Enlightenment Project. As an ideology, and not just as a set of economic arrangements, we find it in the Roman Empire, and again reborn in the Renaissance, until it matures into its fullness in the Enlightenment. With such a pedigree, it is obvious without looking any further at all that it runs counter to periods of traditional integrity in Western civilization. Periods of traditional integrity are notably free of accumulations of capital (thus are they economically and technologically retarded in the capitalist world-view) and indeed

2. As a contemporary example: the tenor and purpose of recent so-called 'free trade' agreements is to prohibit the favour and favouring of neighbors, dissolving local markets into the global pool.

3. Under 'World Civilizations' at http://www.wsu.edu/ffidee/GLOS-SARY/CAPITAL.HTM, 10th August, 2003. To the best of my knowledge, Richard Hooker is in no way a subscriber to a Traditionalist point of view. His 'World Civilization' website purports to give an educational overview of various civilizations and civilization-building ideologies.

traditional societies maintain many devices that actively discourage or disperse pernicious accumulations of capital. Most notably, of course, many traditional cultures—including medieval Christendom—prohibit or curtail usury, but more importantly traditional patterns of work and ownership—from worker's guilds and initiatory vocations to the provision of 'Commons' and the institution of taxes like the Islamic *zakat*—operated against destructive accumulations of capital. Traditional patterns of work are most important of all because traditional cultures are vocational and for the vast majority of souls in any given spiritual climate their work *is* their path. Capitalism, in theory and in practice, is hostile to this and hell-bent on its elimination. This is because, as Hooker again explains in his short history:

> [Under capitalism] productive labor—the human work necessary to produce goods and distribute them—takes the form of wage labor. That is, humans work for wages rather than for product. One of the aspects of wage labor is that the laborer tends not to be invested in the product. Labor also becomes 'efficient', that is, it becomes defined by its 'productivity'; capitalism increases individual productivity through 'the division of labor', which divides productive labor into its smallest components. The result of the division of labor is to lower the value (in terms of skill and wages) of the individual worker.

As Hooker adds:

> this would create immense social problems in Europe and America in the nineteenth and twentieth centuries.

And we might add, it is continuing to create immense social—and spiritual—problems in the 21st century.

What this view of labour amounts to, in short, is a regime of anti-craft wage-slavery. In a regime which 'divides productive labor into its smallest components' the very idea, not to mention the nobility, of the craftsman is atomized and the very notion that work can constitute a *karma yoga* disappears altogether. This is one aspect of capitalism that various Traditionalist writers have exposed with exceptional insight, various articles by

Coomaraswamy being most notable among them, but again without much direct critique of capitalism per se, even though, as Hooker makes clear, the reduction of labour to 'productivity' and the atomization of the craftsman by the 'division of labour' is the very hallmark of the capitalist order. The most conspicuous and obvious and immediate difference between a traditional social order and a capitalist one is in the organization and understanding of labour. And the difference is precisely that all vertical (qualitative and spiritual) aspects of labour are obliterated in the capitalist order and labour becomes a mere quantity, a 'commodity'.

Needless to say, under such a regime the contemplative life has no place whatsoever. Indeed, as Hooker points out, 'the fundamental purpose and meaning of human life [under capitalism] is productive labor' and he notes that this is a position shared by Marxism too. But it is a view of things radically at odds to all traditional ideas of what constitutes the 'fundamental meaning and purpose of life' and is conspicuously aspiritual. It is possible to construe 'productive labour' as a social virtue and even as a spiritual virtue, but in any traditional order the contemplative life, not 'productive labour', is supreme as the human ideal and it is unshakably axiomatic that 'the meaning and purpose of life' is transcendent and to be found beyond the world of objects and the productions of time. Capitalism imposes 'productive labour' as the only worthy human life. This is every bit a grotesque underestimation of the depth and breadth and height of human nature as the atheism of Marx. In capitalism, it arises out of the abstraction of economics. Hooker again:

> Economics, the analysis of the production and distribution of goods, has to be abstracted out of other areas of knowledge. In other words, capitalism as a way of thinking divorces the produc-

tion and distribution of goods from other concerns, such as politics, religion, ethics, etc., and treats production and distribution as independent human endeavors.

This abstraction of economics from all other fields amounts to the radical reduction of all things to their lowest common denominator, and the lowest common denominator among human lives is 'productive labour'. This is a monomaniacal, tyrannical world-view the precise concern of which is to isolate economics and, as it were, hermetically seal it from all other departments of life, and especially those that propose a scale of higher and more noble values, so that it may act to undermine the foundations of all alternatives. It is this aspect of capitalism that makes us suppose that economics has nothing to do with religion or religion with economics. This abstraction and isolation of the economic is functionally important. It means that when a traditional craftsman is stripped of his craft and enslaved to a machine, it is merely an 'economic decision', as if it has nothing to do with his soul. Similarly, on a larger scale, Western industrial powers can colonize traditional cultures economically while insisting 'we don't want to disturb your customs or destroy your beliefs.' The Enlightenment liberalism responded to this with a proclamation of man's 'religious freedom' but this—like the notion of human rights in general—is by way of compensation and is an implicit acknowledgement of the enslavement of modern man.

The Marxist critique of the capitalist order has as one of its principle features a 'base' and 'superstructure' model. The 'base' (and hence the reality in Marx's view) is the economic sphere, and the 'superstructure' of men's beliefs and ideas and thoughts and dreams, and his institutions, his politics, his religion, is mere vapor by comparison, and depends upon and is derived from the base. A Traditionalist reading of such a model is rather that the economic 'base' is of the nature of an abstraction, artificially removed and closed off from every other part of life, including all that the souls and minds of men cherish and all that is indeed Real. And from this removed, abstracted position economics acts as dictator and tyrant. The schism inherent in Marx's model is an image of the self-appointed tyranny of economic life under capitalism—a tyranny whereby all worthy human skills are reduced to merely one, the skill of making a profit. It is worth noting in this—not out of harmony with some of Marx's insights—that by this separation economics places religion upon a leash as well. In fact, by the device of this abstraction, whatever elements of integral tradition remain under early capitalism are co-opted and coerced and by other means turned against Tradition. Thus perversions of integral religions arise which act as agents of capitalism. Arguably, in Traditionalist estimations, most manifestations of Protestantism would come under this, and these days it is impossible to watch late night television without encountering Mammon's shameless and diabolical exploitation of the name of Christ. Thus too capitalism shamelessly exploits and sentimentalizes the institution of the family, while, in fact, its economic 'base' relentlessly corrodes the traditional family and all traditional social relations—the extended family has already been destroyed and the nuclear family is, as its name suggests, atomizing. Yet we find that the fiercest proponents of 'market forces' are the fiercest defenders of 'family values'. In such an environment, our left and right political dichotomy has always been a false antithesis, and in recent times has become glaringly so.

What has remained constant has been the agenda of atomization. For capitalism as a whole, and the capitalist conception of labour, and capitalist abstraction, arises from Enlightenment individualism. Hooker describes the capitalist paradigm thus:

> Capitalism as a way of thinking is fundamentally individualistic, that is, that the individual is the center of capitalist endeavor. This idea draws on all the Enlightenment concepts of individuality: that all individuals are different, that society is composed of individuals who pursue their own interests, that individuals should be free to pursue their own interests (this, in capitalism, is called 'economic freedom'), and that, in a democratic sense, individuals pursuing their own interests will guarantee the interests of society as a whole.

From this we realize that a whole raft of modernist ideologies, including the idols of 'democracy' and 'freedom' are components of the capitalist order, which again emerges as essentially and comprehensively hostile to traditional notions of *kosmos*, man and society.

To give matters proper perspective at this point, we should note that in the fullness of traditional understandings, economics is first and foremost a cosmological science inseparable from a broader cosmology. It has a vertical aspect—namely man's relation to the Creation—and a horizontal aspect—namely man's relation to man, and neither aspect is ever considered to be exempt from integration into the single metaphysical, cosmological and sociological tapestry that is a properly constituted traditional order. Labour, production and distribution cannot be divorced from other concerns, least of all spiritual concerns. In Plato's formulation of the traditional order, his *Republic* of which 'there is a pattern laid up in heaven,' Plato's very starting point is economy: the essence of Justice is that every man does his own true work and the Ideal State is one which permits this to the greatest degree. Moreover, in passages such as the so-called 'Nuptial Number' and in his analysis of the decline of States, it is clear that Plato situates the economy that is the very basis of the truly Just State in a total cosmological framework. Economy is not an area of life that exists in splendid isolation; it is the human cosmology in process. And cosmologically speaking, capitalism is guilty of a false and profoundly anti-traditional teleology. Hooker puts it plainly:

> The economic world view [of Capitalism] treats the economy as if it were mechanical, that is, subject to certain predictable laws. This means that economic behavior can be rationally calculated, and these rational calculations are always future-directed. So, the mechanistic view of the economy leads to an exclusively teleological world picture; capitalism as a manipulation of the 'machine' of the economy is always directed to the future and intentionally regards the past as of no concern.

And Hooker correctly observes:

This, in part, is one of the fundamental origins of modernity, the sense that the cultural present is discontinuous with the past.

This is the essence of anti-traditionalism itself since Tradition, in whatever form it manifests, is inherently 'backwards-looking' since Traditions are temporal developments away from a Revelation—a Tradition, by definition, looks back and clings to its Source. Capitalism has contempt for the past and pursues an idolatrous future from a counterfeit Now. Or, to invoke one of the demons of the traditionalist writers, it is an embodiment of the ideology of *progress*. Hooker has this too:

> Capitalism as a way of thinking is fundamentally based on the Enlightenment idea of progress; the large-scale social goal of unregulated capitalism is to produce wealth, that is, to make the national economy wealthier and more affluent than it normally would be. Therefore, in a concept derived whole-cloth from the idea of progress, the entire structure of capitalism as a way of thinking is built on the idea of 'economic growth'. This economic growth has no prescribed end; the purpose is for nations to grow steadily wealthier.

This world-view further invokes 'materialism' and it is no doubt under that wider rubric that capitalism ought to be considered in any thoroughgoing Traditionalist critique. Hooker defines the capitalist metaphysic: The fundamental unit of meaning in capitalist thought is the object. And at this point he contrasts the capitalist understanding with the traditional or what he calls 'non-capitalist':

> Since capitalism … is fundamentally based on distributing goods—moving goods from one place to another—consumers have no social relation to the people who produce the goods they consume. In non-capitalist societies, such as tribal societies, people have real social relations to the producers of the goods they consume. But when people no longer have social relations with others who make the objects they consume, that means that the only relation they have is with the object itself.

What Hooker does not explore is that in this 'objectification' of the world—which goes hand-in-hand with 'abstraction'—

capitalism rests upon the erroneous identification of the finite
with the absolute, or more exactly, upon the denial of finitude,
for it treats finite things as both indefinite in duration and absolute in meaning. In his devastating critique of modernity, René
Guénon defines the actual malady of the 'Reign of Quantity'—as
opposed to merely its symptoms—by the single word 'extension', by which he means 'cyclic extension' which is, amongst
other things, a descent and a hardening into matter. The 'extension' of capitalism consists not only in the 'horizontal' sense that
it is an ideology that has extended itself across the entire globe,
but in a metaphysical violation that consists exactly in the essential (and from the traditional point of view, dangerous) tendency of capital to preserve or prolong or extend itself. Wealth
arises from nature and then depreciates to nothing as it returns
to nature. Mythologically, Pluto, the god of the Underworld, is
the god of wealth, and for man to become wealthy, wealth must
be removed from the Underworld, but—out of the conditions
inherent in temporality—it will at length surely return it to the
Underworld again. Capital consists in 'withholding' wealth from
nature—withholding Persephone from Pluto, if you like. The
phenomenon of electricity is a good analogy. We extract this
from nature, or in fact from the Earth which is a massive reser-
voir of electricity; thus does electricity yearn to return to Earth
again. We make electricity useful to us by thwarting and withholding and redirecting its urge to return to nature. Thus too
'capital'. This is unavoidable in all but the most primitive econo-
mies, but whereas traditional orders regard it with suspicion—
as it does plundering the Underworld in the first place—the
capitalist celebrates it and makes it the central dynamism of
everything.

The whole metaphysical problem with capital can be
demonstrated simply, if somewhat simplistically as well, in the
following:

If I loan you $1,000 for ten years, at the end of ten years I
expect you to give me back $1,000 (leaving aside interest, etc.)
and in the mean time I say I have 'invested' that $1,000. And I may
perhaps borrow money against that 'invested money' on the

grounds that 'I have $1,000 invested'. The investor here has certainly risked—or gambled[4]—his $1,000 but has not *spent* it.

But where is this money? The person to whom it is loaned must spend it to try to make something of it, but all the while the minimum price at which he can sell his produce must take account of the initial $1,000 which must remain 'on the books' for the duration of the loan. This money that exists 'on the books' and that distorts minimum prices and against which the investor might borrow more is really 'shadow money' since it is in circulation and exists nowhere. Consider the counter example:

I give you $1,000 and at the end of ten years I get an agreed share of whatever you've made of that $1,000 and in the mean time I admit that I've spent (!) my $1,000.

What is happening in the first instance is that capital is treating the $1,000 as wealth that does not depreciate (return to Nature) and in this it denies or thwarts finity. This is the very nature of investment, but when put in these stark terms we can see that it tends to a metaphysical error, namely treating wealth as if it is outside of or immune to time, as a pseudo-absolute. Arguably, this withholding of wealth from nature is the first step towards idolatry and eventually compounds into a forgetfulness regarding the true nature of things. What happens economically is that this shadow money or capital or investment—with the power seemingly to defy Nature and Time—fouls the minimum price at which goods can be sold and the rate of wages that can be paid, messing up the distribution of 'purchasing power' (real wealth) which, as Guénon says in his enthralling chapter on the 'Degradation of Coinage', goes into a long downward spiral. Real wealth declines. Shadow wealth expands. Capital produces and produces and produces but the number of people with real wealth who can buy declines. As we know, this

---

4. The relation between 'investment' and gambling—and of traditional constraints on gambling—requires a separate discussion, but it is obvious that the two things are similar in some essential respects, and just as obvious that in our own times the distinction has become further blurred, many so-called 'investors' in fact gambling—there is no other word for it—on so-called 'markets' which have degenerated into giant electronic casinos.

creates such things as mountains of butter and no one to whom to sell it, and so, finally, you are forced to face the cosmological fact that wealth returns to Nature, and you feed your butter mountain back to the cows—the rational (!) solution. More generally, the mechanism by which wealth returns to Nature in this regime, finally refusing to be thwarted any more, is by war.

Historically, and macroeconomically, so to speak, capitalism's more generalized denial of finity first took the form of geographical or spatial denial: it plundered the New World as if it was endless. When, in the late nineteenth and twentieth century the reality of a finite globe without endless new frontiers arrived, capitalism turned to the pillage of time instead of space and—despite its own future-directed teleology—began devouring the resources that properly belong to future generations. To the present day capitalism is organized around the systematic denial of the fact that the resources of the earth will not last forever and upon the repudiation of any responsibility to conserve resources for future generations, which is to say the outright repudiation of the traditional stewardship ideal. It will be noted that while earlier we observed that capitalism had utter contempt for the past, we now find it devouring the future with a similar contempt—while still straining to maintain the mythology of progress—until we are left only with a counterfeit Now. Modern (and, if you insist on the distinction, post-modern) man is cut off from the past and feels there is no future. There is only a counterfeit present. Counterfeit because it is a parody of, an inversion of, the Divine Present, the sacred Now. Instead of a liberation into the Infinite the counterfeit Now dooms man to an everlasting emptiness, a paradise of shabby consumer goods.[5]

The mission of Traditionalism, of course, is not solely to be critical of modernity but, just as or even more important, to

5. The oft-stated aim of international capitalism today is to maintain approximately 2% economic 'growth' throughout the twenty-first century and thereby transform the entire globe to an American middle-class 'standard of living'. The obvious objection to this utopian vision—quite apart from the economic, political, social and ecological obstacles to realizing it—is that middle-class Americans are among the most restless and least happy people in history.

reiterate and restate the doctrines and values of primordial tradition in and to these times. Again, there is conspicuously lacking in the main Traditionalist writers a thorough account of traditional economics. Here, it is not enough to anathematize capitalism and make a case for its own place in the Traditionalist demonology, it is yet more important to rediscover and restate, in clear and unambiguous terms, what the thinking and practice of the 'best of men in all times and all places' has been on such matters and how traditionally constituted cultures operated on this level. In practice, in a strange *anamnesis*, the errors and manifest horrors and the aspiritual dehumanized atmosphere of the modern era serves to prompt us to hunger for and search out an authentic heritage. Thus the mad, frenzied circulation of trillions of capital transactions around the globe—wealth abstracted now into electronic bytes in cyberspace—can serve to remind us of the traditional virtues of localism. Wall Street can remind us of what a real 'market' was like. High interest rates can certainly serve to remind us that usury was once a sin. We find, that is, that capitalism, as Richard Hooker defined it and as we have discussed it in this paper, is almost exactly *counter*-traditional in its constitution. The Traditional economic order is almost the exact opposite to the capitalist order—although not the false opposite offered by socialism. A traditional economic order is not engineered in any sense, but is organic and consists essentially in complex patterns of obstacle and constraint to prevent accumulated capital reaching the critical mass whereby it snowballs out of control. In the self-flattering, heroic, capitalist, modernist, progressive world-view, the wealth and techno-logical and productive might of modern man—the snowball that is out of control—is the fulfillment of all the struggles and dreams of the human race. In fact, the capitalist order is exactly

---

That affluence does not bring happiness is such a self-evident fact it is truly astonishing that the entire global 'Project' of our times is based on exactly that delusion. Affluent wage slaves are still wage slaves. This is not to mention that such a 'standard of living' is, on the whole, spiritually soporific. The project for global affluence is also a project for global stupor.

what traditional cultures everywhere struggled to avoid like the gates of Hades. Tradition acts as a series of stops against a downward flow that is inevitable and inexorable; this is nowhere clearer than in traditional economic orders which, while often permitting capitalistic elements, nevertheless hem them in on all sides to bind them into a social web that will prevent their becoming destructive. This includes obstructing destructive levels of wealth, for let us remember that wealth is not an unmitigated good in such a view but indeed, while it makes our creaturely life more comfortable, it at the same time makes entry into the Kingdom of Heaven as difficult as a camel passing through a needle's eye.[6] Capital is a necessary evil in a traditional order.[7] In the modern order it is the Holy Grail itself. Plato notes the contrast. In the *Laws* he says that the legislator:

> must watch over the methods by which his citizens acquire and expend their wealth, and have an eye to the presence and absence of justice in the various procedures by which they all contract or dissolve associations with one another.[8]

This is in order to:

> keep [his polity] in subjection to modesty and justice, not to wealth and self-seeking.[9]

This is almost a capsule statement of the traditional case. Traditional man prefers justice and modest means to affluence and entrepreneurs.

An example of traditional economics in action is the institution of child labour. It is a good example because it is a problematic issue in the contemporary context of globalized capitalism and it will hopefully convince the reader that the Traditionalist

---

6. In traditional orders, poverty, not wealth, is deemed the virtue.

7. In traditional Christianity it received much the same degree of toleration as Judaism, the historical relation of which to capital cannot go unmentioned but is a complex issue and would require a separate article.

8. *Laws* 632b. The 'presence and absence of justice' here refers to the preservation of traditional vocations and the justice of the potter being free to pot and the farmer being free to farm.

9. *Laws* 632c.

point of view continues to offer an important perspective on current and on-going crises. We hear stories in the news on a regular basis of corporations (accumulations of capital) running child 'sweat shops' in sympathetic 'Third World' countries desperate for foreign capital. In earlier times, to point to some of the more sinister abuses of the same type, British industrialists once sent children down mines for periods of up to twenty hours at a time and literally worked them to death while gentlemanly spokesmen for British capital defended this practice in Parliament, unashamed to have their words recorded in Hansard. This is the outcome of an atomized, dehumanized view of labour and also illustrates only too well capitalism's contempt for all that comes under the heading of 'family values'. Yet it also illustrates how capitalism, creating and appealing to false dichotomies at every turn, points to child labour as a feature of traditional societies and pretends that it is involved in a quite traditional activity. The truth, however, it is that it is exploiting a traditional institution and at the same time destroying it. For in such 'sweat shops' all qualitative work is abolished and child labour is put to the task of industrialization, which is to say the destruction of traditional vocations. And not long after the capitalists have established their 'sweat shops' in which they are 'only continuing local traditions of child labour' while at the same time bringing 'progress', the liberal do-gooders arrive to condemn it and propose 'modern education' as the solution, removing children from patterns of traditional life in the name of humanitarianism and a different variety of progress.

As it is traditionally conceived, child labour is integral, not exploitative—it is one of the mainstays of traditional economy. Indeed, its importance is difficult to over-estimate. Such a statement is almost beyond the comprehension of modern, liberal sensibilities, yet it is true. In traditional cultures children are often put to a craft at an early age. It is important to the skills of the craft. It is important that the child's young hands and limbs are shaped to the craft from an early age for this allows levels of craftsmanship impossible to attain if you take up the craft once full grown. We now only apply this method of cultivating human skills in sport and in pursuits such as classical music, where we still admit that it is advantageous for a child to start playing either sport or musical instrument from a young age and to be dedicated to it as if to a labour. In the economics of traditional life, what we today admit to be good for children in regard to sport or music—hobbies—applied more generally to work of all kinds. Thus there were long and thorough apprenticeships not only involving the mastering of the direct skills of the craft, but also the customs, lore, rituals and in time the secrets of the craft, namely its most esoteric symbolism, the alchemy of blacksmiths or the Masonry of builders, which explained and incorporated the craftsman into a Divine Work. But it was also important for 'negative' reasons, namely to deliberately 'restrict' or 'restrain' the child from an early age before they developed diverse interests and skills. It concentrated them to a single craft. This is a vital mechanism in vocational societies. It is how, in particular, crafts and trades are passed on from generation to generation in one family. Only a residue remains of this in a capitalist order, and mainly in vocational surnames like 'Smith' or 'Cooper'. Or again, we find a residue of it in the hobby realms of sport and music where we admit—or some of us still do—that, for example, it is quite proper for parents (and/or a school) to restrict a student to the violin from an early age rather than let them dabble with different instruments, mastering none. If we share our child's aspiration that they play high-level tennis, we encourage them to play tennis, not golf. Thus from an early age, by having their choices restricted, a craftsman is crafted to his craft. To do otherwise is to deprive a person of a vocation, which to the traditional mind is cruel and makes children prey to wage slavery and, much akin to it, prostitution. A child that grows up with a smattering of half-developed skills—a retardation of the 'renaissance man'—is the worst possible outcome. In fact, it is relevant to point out here, that there was no such thing as an 'adolescent' before the modern era, nor that pimpled and problematic no-man's land between childhood and adulthood. It is the abolition of child labour—after capitalism has first exploited and industrialized it—that has created the aimless 'teenager' who moreover, in a perverse sociology symptomatic of these profound disturbances, has acted and continues to act, very much like the unemployed, as a type of universal scapegoat in industrial and post-industrial societies. The contemporary debate regarding child labour in the 'Third' or 'Developing' world is falsely and insidiously drawn between exploitative proponents and progressivist abolitionists—it is part of the role of the Traditionalist critique to enunciate clearly what it is that both of these viewpoints are intent on destroying and to underline the fact that what is calling itself 'economic development' and 'progress', not child labour, is the inherent evil. Let us restate it: traditional cultures are vocational and for the vast majority of souls in any given spiritual climate their work *is* their path. Industrialization destroys a spiritual order of which both left and right are ignorant but which is without question the target of the faceless 'system' of which they are the left and right. Wherever it goes, capitalism's lure of 'free choice' systemically excludes only one thing: tradition.

So we see that capitalism in its history, its ideology and its mode of operation, as it continues to function to this day, only with ever greater speed, is fundamentally at odds with everything that the Traditionalist writers espouse, every value they hope to articulate to these times. Hooker's short description of capitalism could hardly make this plainer, but any consideration of it, using whatever definitions, must lead one to the same conclusion. Why then do the Traditionalists fail to identify it as such? In the opinion of the present writer it is proper to envisage capitalism as an 'entity', not just an ensemble of modern - isms, but an coherent and deliberate 'spirit' or 'will', a distinct monster, as we called it at the outset. This is what is lacking in the Traditionalist critique, an appreciation of the way in which capitalism brings the idolatries of the modern world together into a single, willful, all-consuming complex. While the Traditionalists mount the most penetrating and comprehensive critique of modernity as a whole, when it comes to capitalism— although they cover its core ideological aspects—they usually stop short at naming names. Yet here we have a complete world-view at home in the Roman Empire, reborn in the Renaissance, and maturing in the Enlightenment into what is today an all-conquering regime of individualism, progressivism, wage slavery and an uncompromisingly horizontal understanding of man and the *kosmos*. How can it not be named in the Traditionalist demonology? We have already indicated that it is probably best considered as a sub-set to materialism in general, but in the present writer's view, given that it has now emerged as an unrivaled embodiment of materialism and claims the whole world as its domain, and the distractions of socialism are gone, we perhaps have grounds for moving it to yet lower rungs of Hades, if not to Tartarus itself.

In conclusion readers are invited to ponder the following: There is in these troubling and fluid times endless conjecture about the identity of the Great Beast 666 described in the Apocalypse of John, which beast, we are told, 'made it illegal for anyone to buy or sell' without his mark and whose reign is described by John very much like an economic regime, as readers will see if they care to reacquaint themselves with these pivotal prophecies. The only scriptural clue to the identity of this numbered beast is found in the First Book of Kings (and reiterated in Chronicles) where we are told that the number six hundred and sixty-six is *the volume of gold in talents collected by Solomon's Temple in one year.*[10] The number—quite aside from its inherent solar and autumic symbolism—is therefore a symbol of what we might call the Solomonic economy which is a scriptural image of sanctified abundance. When the

10. 1 Kings 10:14.

same number is made the mark of the Beast who stalks the globe in the Latter Days it refers to the inverse of the Solomonic economy, a diabolical and profane prosperity at the expense of, and dedicated to the eradication of, all sanctity. As the markets continue to rally, and the cult of 'growth' expands unabated, invading every corner of the globe, are we not today immersed in exactly such a diabolical and profane prosperity?[11]

11. Note that a traditional critique of capitalism in no way questions that capitalism generates material prosperity; it is efficient, productive, inventive and so forth, and we might even admit that, in the modern capitalist regime, wealth 'trickles down'. But what is the nature of this prosperity, and at what cost?

The Diabolical
Symbolism of the
Automobile

These locusts were like horses...
Revelation 9:5

WITH the possible exception of the television no other item of
modern technology is so pervasive and so ubiquitous and is so
inseparable from the identity of modern man as the automobile.
Modern man has displayed an unstinting and passionate love
affair with this product of his own invention and automobile
transportation has become an unquestionable norm on every
continent on the planet. In affluent societies, mom, dad and
often the kids have cars of their own, while even poor villages in
the 'Third World' or the global 'South' nowadays will usually
depend upon road transport for survival in one way or another,
and a truck or jeep is a hallmark of community progress. It is
possible to conceive of modern life without other technologies,
but the automobile has become so woven into our existence
that it is difficult to imagine life without it. The globe is covered

from one end to the other with trails of asphalt and literally bil-
lions of vehicles traverse them every day. The yellow-brown
palls of exhaust that hang over our cities are the outcome
relentless road journeys requiring countless gallons of fuel. We
wake up in the morning and find there is no milk: without hesi-
tation we climb into our vehicles and drive to the store. We
have established a global automobile culture and it is so central
to who we are and where we are going that we are happy to have
freeways scar our landscapes and are ready and willing to fight
evil and immoral wars to ensure there is cheap petroleum in our
tanks. This whole culture separates modern life from all that
came before. Unlike our ancestors, we have all climbed into
automobiles and traveled roads from one place to another at
speeds unimaginable by horse, and many of us have spent long
hours in automobiles and indeed some of us have spent a good
portion of our lives in them. They are unavoidable: a fact of life.

Demiurgic

In a traditional perspective it is clear that all God-created things
are part of a symbolic order and have their own inherent symbolism. Indeed, Creation is in whole and in its parts a manifestation
of the Divine and, as far as man is concerned, a revelation consisting of 'signs' for him to understand. What though of man-created
things? Man is the microcosmic encapsulation of the cosmic
order and to this extent—his comprehensiveness—he exercises a
god-like power by which he can 'create' objects that are seemingly as real and as integral as 'natural' or God-created objects.
What is the status of these things? Are these part of a symbolic
order with a symbolism of their own? The answer to this question is yes, certainly, for there is nothing man can do that is not
symbolic at some level since this is the very nature of the Creation, and his houses and furniture, clothes and effects are all
symbolic of something. And this necessarily extends to his
machines of both high and low technologies. When man makes
he exercises a God-like demiurgic power. In traditional orders it

is understood that measures must be taken to sanctify these powers and to exercise them in ways befitting their sanctity and situ-
ating them within the total order of symbols. The Greek
tradition gives us a good example. To the Greeks the *kosmos* was
understood to be a crafted object, as the word *kosmos* itself
implies, and was made from primal materials by a Craftsman god.
This is the cosmological God, the lower aspect of the Divine
Being that engages with creation, deigning to dirty His hands in
labour, so to speak, unlike the higher aspects of Divinity which
remain aloof from the Creation. In the Greek pantheon the
demiurge was the smith god Hephaestus, the Hellenized adaptation of the Egyptian potter god Ptah, the lame-legged Olympian
who tarried in his workshop all day, manufacturing trinkets and
gadgets and mending objects for the other gods who, in the main,
found his antics enormously amusing. The human smith in the
Greek order reveres this Olympian deity as immortal exemplar,
but more importantly the whole art of the smith is understood in
terms of *mimesis*—imitation—of the divine model. It follows that
one observes the action of the exemplar in the divinely crafted
objects of nature, and so not nature but the *action of the exemplar in
nature* serves as the basis for *mimesis*. For example, the white-hot
flux of metals that occurs in a volcano is terrestrial evidence of
the applied arts of the Olympian smith, Hephaestus (Vulcan). The
human smith, in a context of reverence for and awe of the Olympian model, will imitate the volcano and the arts of the god in his
furnace. Again, he does not imitate nature; he imitates the *god in
nature*. By this means the productions of the forge attain a type of
existential legitimacy: such productions have a God-approved
legitimacy as much as trees and rocks. There is, in this sense, no
distinction to be made between God-made and man-made
objects for the man-made object is the product of a sanctified *participation* in the work of the God; the craftsman works with or for
the God, imitating the way the God works (in nature, his handi-
work) just as an apprentice will imitate the ways of the Master.

At the same time, however, there can be no escaping the fact
that a man-made object, though it may be made by participation
in a divine work and in this way woven into Creation, is never-

theless a 'creation' of a lesser order, for the simple reason that man is man and not God, and the truth of the matter is that he can create nothing that is really *new*, all his so-called 'creations' being reworked from existing materials, namely the divinely crafted *kosmos*. In all his productions man is recycling materials that have already been through the primal forge of the divine craftsman. Even the most sophisticated high-tech engineer is really like a backyard inventor recycling junk and spare parts from nature. There is a necessary sense, therefore, in which all human production is secondary and a man-made object is always a remove from natural objects. When we say that God crafted the *kosmos* from pre-existing materials we only do so for convenience, just as we distinguish a demiurge from a higher deity only for convenience. But in the case of man it is literally the case that he must start with pre-existing materials, so while he can 'create' demiurgically from a *materia* he cannot, like God, create *ex nihilo*. God is not merely Demiurge but in so far as man's powers over nature are God-like those powers are *demiurgic* and *can be nothing more*, for man cannot create from nothing and even his finest productions are, at best, recycled goods. There is therefore something inherently flawed in human productions vis-a-vis natural objects. They can only ever be *like* natural objects but never *be* natural objects by having the same relation to the Principle. Man's productions are one step removed from the true *prima materia* which—while we speak metaphorically of God as forming the Creation from a pre-existing material (His demiurgic aspect)—is actually the Nothing of *ex nihilo* (His higher aspect). *Mimesis* is also not without its inherent moral and spiritual dangers, for in the exercise of man's demiurgic powers that are a consequence of his microcosmic internalization of the forces of the *kosmos*, the distinction we have just made is liable to be overlooked and man soon starts to think of himself as god-like in a fuller and inappropriate sense and his 'creations' as primary. Again, man creates nothing. He recycles.

In a traditional order, as we said, we find that technological innovations are carefully sacralized and integrated into the continuum of tradition, even if radical adjustments need to be made to weave the new technology into the total symbolic framework that is the matrix of such a society. Sacralization, though, always consists of ways and means of ensuring that the inherent limitations and dangers of man's productions are understood. The way that the plough was woven into traditional symbolism illustrates this well, to cite one example, or, more relevant, the way the metalworking technologies were given the symbolic matrix of alchemy. The traditional blacksmith knew well that his materials are already crafted objects (crafted by the Divine Blacksmith) and whatever he makes of them can never be *pristine* because he can never *be* God but only a co-worker to God, and then in only one of His aspects. It happens though, because of cyclic degeneration, and because it is in the nature of technology for one innovation to suggest another and then another and so on, that technological developments inevitably out-pace every effort to integrate them into the symbolic frame-work, and improvements and modifications in technology call for such complex adjustments to traditional symbolism that, eventually, new technologies evade sacralization and the traditional constraints and balances cease to be effective or disappear altogether. The obvious example to be cited here is the invention of the printing press which technology—the technology of mass literacy—could not be integrated into what remained of a traditional order in Europe, as we see in the desperately heavy-handed and clumsy devices by which the Inquisition and the Index attempted to enforce some degree of orthodox restraint, and in the fact that they failed so comprehensively to prevent the Protestant Revolt from using the technology to rupture Christendom. Here is a technology that 'started a revolution' as the historians say, in this case a decisive rupture from the unified spiritual ideal of the Middle Ages and a catastrophic breaking point for the Christian order. By this time in European history we are already aware that the fabric of tradition is tattered and that new technologies will not be woven into the fabric but will tear new holes. The clock, so long as it had a round face and two hands in a cosmological soli-lunar order of symbols, could find some symbolic integrity, but little compared to the times and seasons kept before the mechanical regularity of clock-time. Needless to say, by the time the automobile was invented there was no prospect whatsoever that it could be integrated and so to speak *neutralized* within a matrix of traditional symbology because the Western tradition was in mere threads and European man's pursuit of his demiurgic delusions was well advanced. There can be no question, therefore, that the automobile and all its associated technology is diabolical—or in Greek terms, Promethean—because whatever demiurgic technology has not been sacralized is so.

This does not prevent us from examining such a technology as the automobile from within the framework of a traditional symbology, however. As modern man's most prolific 'creation' it will surely reveal something significant about the predicament in which he finds himself. From this point of view, there are two peculiar and unusual things about the automobile that require attention: the fact that it has the appearance of being *self-moving* and the fact that its cabin, into which human beings climb, forms a separate *space* from its environment. The automobile is, by definition, a self-moving machine, as the term 'horseless carriage' suggests—its whole construction gives the illusion of it being self-moving—and in its typical form it is like a capsule, an interior space. In both these cases let us note that these are characteristics of *animals*. Animals, by definition, are self-moving (*autozoon* in Greek) creatures. The automobile mimics this characteristic. And animals (and man) are 'capsules' in that they form microcosmic interior spaces with internalized organs and an internal environment distinct from the external environment. The car is like this too. The interior is a separate space, with exterior sounds muffled, and increasingly, especially in contemporary vehicles, a whole world of gadgets within, every comfort of home on board. So in these respects the car is like an animal: most obviously like a *quadruped*. The fact that it replaced the horse is enough to make this plain: it is, amongst other things, a metal horse. An understanding of this is the starting point of any symbolic consideration. The automobile is, first and foremost, an artificial beast. It has four legs in four tyres and eyes and mouth in lights and grill and its power is still measured in horse-power. Modern man's obsession with the automobile is a direct extension of an earlier preoccupation with the horse. But unlike travel on a horse, the automobile traveler *climbs inside* the vehicle and so travels within the beast, so to speak, occupying the micro-cosmic world of the cabin. In this respect the beast has been combined with the cart or carriage it once pulled. The mytho-logical parallel with the Trojan Horse must be pointed out here; although it is not self-moving it is nevertheless a foreshadowing of the automobile—an artificial horse on wheels into the inte-rior space of which men climb and sit. The Trojan Horse, of course, is an emblem of the sacrilegious sack of Troy and so by extension an emblem of cyclic decline and not in any sense a felicitous symbol. It brought ruin to Ilium and the curse of blas-phemous plunder to the Greeks. It is the preeminent emblem of *hubris*.

Jurassic Technology

The image of the Trojan horse allows us a further imagery, for it was a gargantuan horse and in some ways might remind modern man of the *dinosaurs* of the fossil record. The analogy becomes more apt when we remember that the automobile is fuelled by the processed residues of former aeons now compacted into subterranean lakes of crude oil. The horseless carriage does not really look like a horse, at least of the modern type, but more like some squat, flat-faced, prehistoric ancestor of the horse with a plated protective skin. There is indeed something dino-saur-like about many larger road transport vehicles; it takes little imagination to see this if one is standing on a roadside at night as trains of large transport vehicles roar and rumble by. If the auto-mobile is like some artificial beast from a former aeon, the truck and lorry are like large multi-legged prehistoric monsters. It is characteristic of the later stages of a cosmic cycle for men to plunder the remains of the earlier stages, bringing into circula-tion with the plunder the psychic residues of those earlier stages. The technology of the automobile is a example of this. To fuel this technology modern man removes from the earth the volatile residuum of former ages and makes from it food for his metal horse which, morphologically, is not an improvement on the horse in any sense but more a reversion to the grotesque quadruped forms of the prehistoric past. There is a strongly 'Jurassic' motif in this technology that must be noted as one of its most significant characteristics. What manner of quadruped is the car? Its form is clearly not like that of existing animals, even though it takes its departure from its immediate forerunner, the horse. Where do we find quadrupeds large enough that men could conceivably sit inside them? To find resemblances we need to look at many dinosaur life-forms: just as the Trojan Horse was titanesque, so the automobile is the return of a quadrupedal morphology from a distant era. There is no escaping the implication that this technology is therefore inherently monstrous and unleashes malignant, chthonic forces held in check until modern man released them from the earth.

Internal Combustion

In alchemical symbolism this is expressed as a *dragon* motif. The locomotive has an obvious resemblance to the classical dragon—especially when locomotives were stream-driven—but so too does a road-train, and there is something dragon-like about the entire automotive technology. The steam-driven locomotive was literally fire-breathing: in automobile technology there is still the exhaust to suggest fire-breathing but more particularly we are reminded of the way dragons carry fire within their bellies by the *internal* combustion engine, the very *internalness* of the combustion being the parallel. It is not an accident of symbolism that crude oil is called 'black gold'. The lakes of oil under the earth are, in fact, the residuum of the aurumic humus of Edenic times, the physical residue of the gardens of the

Golden Age, and so are in that sense the treasure of the alchemical dragon. Modern man has stolen this treasure and unleashed the dragon. The petroleum sciences are, then, quite exactly a *counter-alchemy*, a diabolical alchemy that hastens the onset of cyclic dissolution rather than preparing the way for the new cycle. In recent decades it has become obvious that this technology entails transposing the heat and carbon contained in these lakes of oil from the earth into the atmosphere and that this is likely to have a profound impact on the polar structures of the earth and so constitutes a transformation of global, geological proportions. In the long run this must have an impact upon the entire balance of the terrestrial system and perhaps even upon the earth's axial balance and magnetic polarities and such like. We are belatedly beginning to realize that this—equal to the atom bomb—looms as the greatest threat to our own existence we have yet engineered. Now we face the dragon in, amongst other things, the storms and monsoons, tidal waves and wild perversities of weather that follow from emptying chthonic residues into the atmosphere. This is the full context in which the humble, everyday automobile is dragon-like. In European medieval dragon mythology, also let us note, the dragon is confronted by a knight *in armor*. The metal, protective skin of the medieval knight prefigures the same in the automobile. In the automobile we find the motif of dragon—with its internal fire—and the motif of the knight's armor combined. Much medieval dragon mythology concerned technological triumph and prepared the spiritual conditions for industrialism. Many of the most basic mechanisms used in automobile technology, such as crankshafts, are actually medieval in origin. The metal-plated knight slew the dragon, stole its treasure, and acquired its powers, specifically the power of internal combustion.

Exoskeleton

The extension of metal armor plating from man to vehicle is most suggestive of the tank, the military adaptation of the automobile. The Trojan Horse was tank-like too, a military device, and here we must remember the intimate connection between this type of technology and military motivation, the human drive to find new and better ways to bring death and mayhem to the earth. Most technological 'advances' are of military origin, not the outcome of humanitarian sentiment, another instance of the way they are sponsored by the destructive forces of cyclic decline. In the case of the automobile we can see ancient precedents in such military innovations as the Roman's famed 'turtle' formation, where groups of foot soldiers would lock shields on all sides and overhead and move into battle as a single, impenetrable 'vehicle' that, since the legs of the troops inside the formation were hidden by shields, appeared to be self-moving. The parallel with the turtle in this case is again a comparison with an animal, this time with the emphasis on the idea of a *protective shell*. Aside from a mammalian quadruped symbolism, devolved from the horse, the automobile has an obvious 'turtleness' in this respect, and in fact the turtle is a quite appropriate and traditional symbol for the microcosmicness of the automobile's cabin.

Even more appropriate, though,—since the turtle is slow—is the same idea expressed in other creatures with exoskeletons, like insects. Frithjof Schuon observed that there is something profoundly insect-like about the conditions of modern living and he compared our sprawling cities to vast hives of frenetic insect activity. In this analogy our automobiles are very much the exoskeletal insects that scurry to and fro throughout our ever-swarming urban hives. A modern city seen from the air is like an ant heap. The way the automobile has devoured the globe is truly comparable to a Biblical plague of locusts. This analogy is particularly evident in the famous German designed *volkswagon* which indeed looks insect-like or locust-like and is popularly called a 'beetle' or a 'bug'. In the Bible, in fact, we find a peculiar conjunction of the symbols we have discussed: insect and horse. In John's Revelation we are told there will be *locusts*—with the powers of scorpions—which are *like horses* and they are even said to be covered in iron body armor and to make a din, and their appearance is accompanied by the emptying of the Abyss, the smoke of which chokes the atmosphere and obscures the Sun. Another appropriate symbol, with important astrological resonances, is the crab, the zodiacal image of enclosed microcosmicness which is essentially interchangeable with the turtle or tortoise as a symbol but is fast, not slow. The jerky start-stop, scurry-stop, scuttling of the crab is very much like urban automobile travel and even the indirectness of the crab's propulsion has a parallel in the quite peculiar (even counter-intuitive) shift of energy from the motions of the engine's pistons to the turning of wheels in which the 'drive' is indirect.

Crustacean

The implication of this exoskeletal technology for man himself is, of course, that he is becoming a *crustacean* as he lives more and more of his life—from conception to death—in the protective shell of his automobile. Increasingly he feels more at home within this shell than he does stepping out into the fresh air. When he feels like communing with nature he drives to a vantage point to sit in the car and watch the sun set, listening to a CD and enjoying drive-thru food and drive-thru beer purchased with drive-thru money. New technological endeavors are devoted to finding more and more ways to enable the motorist to conduct more and more of his life without once stepping out of his car. Indeed, technological visionaries suppose that soon motorists might be physically connected to their automobiles by way of biotechnological devices and really become part of the vehicle. Naturally, the more man adopts this exoskeleton the further his existence is removed from the pristine craftwork of nature—he is further abstracted from reality—and also the more his inherent bodily powers atrophy.

Traditional man in whatever era walked a great deal in their lives. Traditional life is local but it also insists on pilgrimage and, even with horse travel, walking was the normal mode of locomotion. It is the uniquely human mode of locomotion that cannot be compared to the gait of any other creature, contemporary or Jurassic. The left-right alternation of walking, moreover, is integral to the human form and is directly analogous to the two halves of the brain so that, in quite a biological as well as symbolic sense, walking is a parallel to the basic operations of thinking. Modern man walks very little. He drives. This is to say he sits. He spends more and more of his time sitting within the metal and glass protective shell of his automobile. Walking has been reduced to a few movements of the feet on pedals and in clutchless cars to just the accelerator. It is instead the upper body, the head and arms and hands with which ones drives. It is a particularly cerebral, head-focused activity compared to walking. The thinking that accompanies driving is a-physical and abstract. Consequently in this, as in other ways, modern man is being reduced physically and hardened mentally. He needs the protective shell because he is becoming more puny and more vulnerable in himself. This is the tragic paradox of technological man— the more gargantuan his technology becomes the more he is himself diminished as a creature. He is dwarfed by his own giants. He is the little man in the big machine, the Wizard of Oz. Modern mythology projects an image of this in the typical characterization of the technologically advanced 'aliens' or creatures from outer space—pale, shrunken creatures with atrophied limbs and huge heads. Man empties himself into his technology. Bit by bit he replaces his internal faculties with exterior devices. This is the way of the cosmic cycle. Man is *most* microcosmic at the beginning of a cycle. He loses this integrity, however, and throughout the cycle *his microcosmic powers are emptied back into the macrocosm.*

Every technological advance injures some aspect of man's primeval integrity. Man conquers nature by emptying himself. The conquest of nature is thus profoundly self-defeating. The discovery of fire weakened man's internal fire. The invention of shoes did injury to his feet. We are currently exteriorizing the human nervous system into computers and the immune system into vaccines. More and more the human microcosm loses its integrity vis-a-vis the macrocosm. Cyclic decline is an *exteriorization:*

the exoskeletal automobile is an image of this in our times. The zodiacal symbol mentioned earlier, the crab, calls for more comment here. In the conventions of modern Western astrology the zodiac begins at Aries and so Aries corresponds to the head in the human body. But an earlier symbolic order has the zodiac beginning at Cancer with that sign corresponding to the head. In this symbolism the crab is analogous to the exoskeletal human cranium. In the symbolism of the greater cycles Cancer is the primal age and the crab is an image of the microcosmic completeness of primordial man. But the beginning is also the end, and so there is a zodiacal symbolism underlying modern man's metamorphosis into crustaceans: as the end of the cycle approaches and man has emptied his internal powers into his own productions, exteriorizing them, the primal symbolic is reversed parodically so that modern man in his automobile is a counter-image of the primordial microcosmic integrity of the men at the start of the cycle.

Hephaestus' Workshop

Returning to a Greek vocabulary, the sitting posture normal while driving, and the consequent decline of the uprightness of walking, and the idea that a life of this inevitably damages the primal integrity of the human form and its capacities, recalls the lameness of the demiurge, Hephaestus. It is commonly supposed that Hephaestus was made lame by his fall from Olympus to Lemnos, but in fact he was lame from birth and so a defective deity among the Olympians. Automobile technology is very precisely Hephaestean in this regard. Hephaestus is lame: his lower body has atrophied. He has an exaggerated upper body but spindly legs. He hobbles about playing with his gadgets and inventions. The motorist—his lower body irrelevant—is an Hephaestean being, symbolically lame. There is, in fact, in the mythology of Hephaestus recorded in Homer's account of the Trojan War, an uncanny foreshadowing of the self-moving vehicle presented as an Hephaestean device. In the smith god's workshop, we are told, there are a set of metallic stools, forged from the god's furnace, that scuttle to and fro the assemblies of the Gods all of their own accord, like self-moving and intelligent creatures. In the same passage we also meet a group of 'golden maidens' crafted of metal but who are nevertheless self-moving and endowed with *nous*, in what other writers have correctly observed to be a prefiguring of the modern robot. Modern technology has realized the contraptions of the Hephaestus' workshop and the automobile is the realization of his fabulous self-moving seats. Hephaestus is a binding god too, and we note the way the traveler is bound into the cabin of the automobile by belts and straps. But there is no sense in which modern technology participates in a sacralized Hephaesteanism: rather the technology has now been stolen from the god who himself has disappeared in man's demiurgic intoxication and plunder of the earth. Inevitably, there must be demonic and diabolical forces associated with such a technology and indeed we see aspects of this in the way certain people develop obsessions with cars, in the phenomenon of 'road rage' and of 'speed demons', mild mannered people who are aggressive, maniacal drivers, and in the others ways people manifest forms of psychic possession regarding cars. Every diabolical technology collects human victims whose lives are overtaken by the technology. Television is the obvious example. It impacts upon most of us, but some people it utterly absorbs and in effect destroys. The automobile is the same. The nature of the possession might be described as microcosmic collapse. Without the machine there is nothing left. Without his car modern man is stranded and cold and exposed. As the poet said, *the center cannot hold*. In the End of Days men become like rootless spinifex in a frenzy of pointless transportation from A to B and back again and live their crustacean-like lives as a fitful journey to nowhere looking at reality through a windscreen. Traditional symbolisms provide ways to understand the diabolical nature of these things.

Notes on the *Parmenides*

THE *Parmenides* and the *Timaeus* are related dialogues. Both are set on the Panathenaea (the festival of the Goddess). Both dialogues are relevant to that festival and to the gods and rites of that festival. (The *Timaeus* is the dialogue of *cosmology*, the *Parmenides* of *metaphysics*.)

According to the *Parmenides*, Plato feels that the appropriation of Homer by Athena (on the Panathenaea) has introduced an error into the Acropolis cult—namely the error (from the Odyssey) that Athena and Poseidon are rivals, that they quarrel. Plato sees this as introducing the error of dualism into the city.

Plato teaches that the gods do not quarrel; this is the whole point of Solon's unwritten dialogue (the *Atlanticus*)—it was to have been a type of Athenian counter-*Odyssey* in which Athena and Poseidon are not rivals but complements. The war between the Athenians and Atlanteans is not a result of divine rivalry but is caused by mortal error when the 'immortal blood' of the gods is reduced (and autochthony replaced with animal reproduction).

The *Parmenides* is a dialogue of Poseidon (while the *Timaeus* is of Athena). In *Parmenides* 137a, Antiphon, after fully absorbing (by heart) the Parmenidean metaphysic, has retired to tending horses. 'I feel', he says, 'like the old race horse in the Ibycus who trembles at the start of the chariot race.... I am frightened at setting out to traverse so vast and hazardous a sea.' (This may in fact be an allusion to a poem by Parmenides about a chariot ride.) Antiphon is shown as instructing a blacksmith in how to make a bit for a horse. Here the smith = Hephaistos, and the horses and the sea = Poseidon. This is as clear as it gets!

The *Parmenides* is set on the Pananthenaea in order to illuminate one of the metaphysical secrets of the festival, namely the paradox of non-duality. The dialogue commences with a para-dox of Zeno; the problem with Socrates' account of the Ideas is that it cannot withstand paradox.

The *Parmenides* dialogue of Plato—infamous in its obscurity—is essentially a metaphysical account of the dictum 'the gods don't quarrel! (or have rivals)'. Homer has admitted an unlawful opposition; Parmenides stands against the linear either/or, yes/no, this/that (Boolean) Socrates, and expounds the para-doxes of non-dualism: whether or not there is a one, both that one and the others alike are and are not and appear and do not appear to be, etc. In the *Parmenides*, Plato comes very close to the Indian Vedanta in Greek mode.

The teaching of Plato's *Parmenides* (at the Panathenaea) reveals the true meaning of the relationship between Athena and Posei-don (gods of the city and the Acropolis cultus). Athena and Poseidon are mythological renderings of the metaphysical hypostases Essence and Substance. Note the ship symbolism of the Panathenaea. Athena = vertical (mast, sail, wind, air); Posei-don = horizontal (keel, sea, water).

What Plato calls the 'philosophical nature' is (mythologically) the 'immortal blood' of the primal autochthons. (Aristocracy is a prolongation of autochthony; all royal lines go back to autoch-thons.) However, the immortal blood of the Atlanteans eventu-ally becomes diluted with mortal blood; this is the reason for the fall of Atlantis.

PART THREE

The Secret of Violets

Discussions on
Traditional Cosmology

The man who lives in a natural environment has a fair chance
of reaching out to something beyond nature; but he who lives
in an artificial environment has done well if he succeeds
simply in remaining human.
Gai Eaton, *King of the Castle*

What is the main difference between the modern scientific order and
a traditional cosmology?

THE distinction between the traditional sciences and the modern scientific world-view—or the industrial world-view—is the
distinction between musical and harmonic proportions on the
one hand and arithmetic, serial numeration on the other. The
traditional mind is inherently musical—its habit is to view the
phenomena of nature in terms of musical relations and harmonies. When we say that there is a parallel or an analogy between
one thing and another, or that one thing is a symbol of another,

there is an implicit appeal to harmonic proportion. This is most obvious in such traditional doctrines as the Pythagorean 'music of the spheres' whereby the *kosmos* is understood as intrinsically musical. If there is a correlation, a parallel or an analogy between, say, a flower and a star it is because they share a common order, or are expressions of a common 'note', a common 'vibration'. The modern mind, the industrial mind, senses no such correlation because the modern scientific order exposes only the arithmetic relations of the phenomena of nature. There is no sense of a cosmic harmony in the industrial mentality. The modern enterprise involves only serial, arithmetic relations. The modern mind thinks in sequences of quantities. The traditional mind thinks in musical proportions. It is the difference between the number sequence (1, 2, 3, 4, 5 . . .) and the musical scale (A, B, C, D, E, F, G, A). The number sequence proceeds by the addition of quantities. The musical scale proceeds by proportions. Plato's formula from the *Timaeus* is musical in nature. Fire is to Air as Air is to Water as Water is to Earth. It is a set of proportions. Proportionate relations. Numbers are not like this but musical notes are. Proportion is musical, not arithmetic. In Plato's cosmological science—in all traditional cosmologies—he is looking for the *proportionate* connections between the things of nature. Modern science—industrial man—looks for the *arithmetical* connections. The advantage of seeking the arithmetic connections is that it gives you control over the gross mechanics of nature, the levers and pulleys, quantifications, but this at the expense of its inner harmonies. This at least is a helpful way to understand the difference.

Is there one governing idea in traditional cosmologies?

One of the governing ideas in ancient and medieval cosmology, east and west, all cosmological sciences before the modern era, was that the human being encapsulates the *kosmos*—the human being is a microcosm (small *kosmos*) that is a copy of the macrocosm (greater *kosmos*). This is both implicit and explicit throughout ancient and medieval sources. It is an essential idea. It is fundamental to all spirituality, all religious thinking. It is

impossible to understand Plato and other traditional philosophers unless we grasp this notion. Plato uses this idea all the
time. Most notably, in the *Republic*, Socrates suggests that the
best way to understand Justice in a state is to understand it in a
man. Man and State are therefore parallel. In the *Timaeus*, Plato's
dialogue on nature, Plato adds that both State and Man are also
parallel to kosmos. This is typical of traditional cosmological
thinking. Traditional man thinks in parallelisms. The whole is
implicit in the parts. The whole *kosmos* is implicit in man. Man is
a microcosm of the macrocosm. In some ways all traditional
cosmological thinking explores this governing idea. And without this governing idea you really cannot understand religion. If
you try to reconfigure religion to a modern, industrial cosmology that is devoid of this idea, and devoid of analogy, stripped of
parallelisms, you end up with bastard superstitions. The modern
is the *death of analogy*. Specifically, there is no analogy between
Earth and Heaven, no analogy between Man and *kosmos*. Modernity is the breaking down of a universe of analogies.

And replacing those analogies with what?

With perversions of traditional doctrines. Whatever perversions
will permit the unleashing of the forces of production. Whatever
perversions will facilitate control and power. Power—power is
the key word of modernity. And in so-called post-modern thinking it is laid bare—there is nothing else. Typically, though, the
modern industrial mode replaces traditional doctrines with perversions of the same. Thus, for example, evolutionism is a perversion, a perverse version of the macrocosm-microcosm doctrine.
Darwinian science didn't invent the idea. It is implicit in the very
ancient hermetic doctrine that the microcosm reflects the macrocosm. Which is really the doctrine of the earthborn. There is
nothing remarkable or even supernatural about it. We are the
product of our *kosmos*. And we take the shape and functions of
our environment. If, for example, the Sun and Moon and Earth
were rectangular instead of circular then I think the human head
would be rectangular and the pupil of the human eye would be
rectangular, whereas they are circular now because the Sun and

the Moon and Earth are circular—spherical, of course—and we and our organs reflect essential features in the environment from which we came. So microcosm reflects macrocosm. And the Earth reflects the Heavens. This is not in any way a mystical or supernatural or anti-natural, non-natural, idea. A field of wild-flowers reflects the starry sky. The wildflower is star shaped. This is not astrology. It is just saying that the things we see around us are a product of their environment too. We need not suppose any supernatural—or anti-natural or non-natural—force. But *circular environments create circular forms*, that's all. The theory of evolution says that, implies that, too, doesn't it? Circular forms will be more fitted to an environment of circular forms? Yet in modern empiricism, so-called, we cannot make analogies. We cannot observe that A is *like* B because both are spherical, or because both are star-shaped, even though this is obvious to our senses. Any science based on our senses must be a science of analogies. The analogies and parallels in the *kosmos* scream out to be noticed. Industrial science isn't really a science of observation. It refuses to see harmonies and proportions. They are everywhere to be seen, but it *refuses*. It refuses to hear the music of the *kosmos*. It is only interested in the levers and pulleys of nature, the crude mechanics, that allow *control* and bestow *power*.

> *But hasn't man always been searching for those levers and pulleys and in modern times, with modern technology, we found them? Weren't the traditional cosmologies looking at the wrong connections?*

In the progressive historical narrative, the narrative that modern man flatters himself with, all history was a long struggle for this present age. So by that reckoning we are the envy of history. We've achieved what man has always struggled to achieve. But in traditional accounts there is another narrative. Traditional narratives tell us something quite different. On the contrary, they suggest that the modern era is what human beings have always tried to *avoid*. It is the opposite to the narrative of progress. In mythologies and doctrines, traditional doctrines, throughout the world, we have accounts of how people want to *avoid* the technological age. It is portrayed as the Iron Age, the Ferric Age, the age of corruption and destruction, in contrast to the Golden Age. In some ancient accounts they describe the modern world perfectly, but they don't regard it as the summit of human achievement. Although they do, all the same, see it as *inevitable.* The Iron Age must come. It had to be. But traditions, traditional orders, are like *dams holding back the flood.* It is much more accurate to portray all human endeavor as attempts to forestall the modern era, the Iron Age. Our religions, our traditions—they wanted to *prevent* the modern world. But in the end, it came like a flood and nothing could stop it. You find both narratives in traditional sources, in fact. In Plato you find both. In Plato you have the progressive narrative, how man began living in caves and slowly eked out an existence, conquering nature, taming fire. But also in Plato—at the heart of Plato—you find the Golden Age mythology whereby history is a process of decline.

Which narrative is true?

Both. But they refer to different things. You can't say that the musical scale is true and the arithmetic sequence is false. Both are true. Both are real. But they are different things. Between them there is a process of internalization and externalization. What was once internal becomes external and vice versa. This is what has happened in the descent into modernity. Industrial man might be externally rich—rich in science and medicine and cars and houses and such—but he has become internally empty. He has satellites and mobile phones *but no one to talk to and nothing to say.* We are in the process of externalizing ourselves. By technology. So that, for instance, eventually the test-tube replaces the womb. The computer replaces the nervous system. Our inner resources are being replaced by external means.

So technology replaces our innate capacities?

This began long, long ago. Fire and clothing replaced our internal warmth. The book replaced our memory. Writing, the technology of writing, externalized our memories. We no longer had to remember things internally—we wrote them down and kept them externally. Plato relates the myth of Thoth, the Egyptian god of writing. When Thoth teaches writing to man it is touted as a great advance, but of course the other side of this supposedly great advance is that it retards the human memory. Modern man has a puny memory compared to ancient man. In this regard a modern intellectual is a pigmy, a simpleton, compared to an aboriginal elder or a Lakota chief from a pre-literate culture. The modern intellectual has his books but his memory is miniscule. He knows nothing by heart. In the so-called march of history inner capacities are externalized. Technological man empties himself into external applications, so to speak. It is an emptying. A casting outwards from within. Industrial man empties himself into the outer world. The microcosm is emptied into the macrocosm. And similarly what was once external to man has been internalized. It is a reciprocal process. Traditional cosmologies describe these alternations. There is a very important myth in Plato's dialogue called the *Statesman* or *Politicus*. It is all about this alternating process. Traditional cosmologies anticipate the modern age very exactly, describe it in detail. But it is not something of which man has always dreamed. That is modernity's myth about itself. It is probably truer to say it is more a nightmare man has always avoided.

What do you think is the main feature of modernity?

Quantification. The arithmetic scale is a scale of quantities. 1, 2, 3—these are quantities. The keynote of the entire modern enterprise is quantification. Modernity is the reign of quantity, to use the name given to it by the greatest master of symbolism and metaphysics in the modern era, the Frenchman, René Guénon. The reign of quantity. Because so-called 'Enlightenment empiricism'—its guiding characteristic—is that it will only permit those connections in the natural order that can be quantified. Size. Speed. Volume. Weight. The modern sciences are sciences of quantification. The traditional sciences are about qualities—proportions are not quantities—and not about raw quantities. The quantities—size, speed, volume, weight and so on—these give you access to the levers and pulleys, the mechanisms of push and pull, in the physical world. That is all that industrial man cares about. Modernity is not very subtle. All it understands is power. The nuclear bomb is really no great achievement. It is rudimentary engineering and undergraduate physics. We found nature's levers and pulleys. The cogs and chains of the atomic world. And it gives us power. It doesn't give us wisdom or self-understanding. Just power. Pure force. Lots of it. Quantity. And similarly, the economic order of industrialism—capitalism—is about *quantifying* the world. Traditional social orders always avoided, prevented, accumulations of capital. The modern order is built upon accumulations of capital. It was always a temptation that man would do this. Eventually the temptation was just too much.

Why do you use Plato as your source for these ideas?

Because Plato is the great spokesman for traditional ideas in the West. Plato is the great exponent of tradition, *the* Tradition, in Western ideas. The great fountainhead of traditional cosmology in the West is Plato's dialogue the *Timaeus*. And throughout Plato's works... there is a further sub-stratum of Plato. Very important. The Platonic myths. The myth of alternating cycles of time, backwards and forwards, in the *Statesman*. The myth of Er in the *Republic*. The Nuptial Number. The Protagoras myth. The tale of Atlantis. The simile of the Cave. These are ancient tales embedded in Plato. In Plato we find a core of so-called myths, myths about the cycles of time and about *autochthony*— these are myths, stories, motifs, found across a great range of cultures in many eras. Plato connects Western ideas—philosophy—with the *sophia perennis*, the perennial wisdom.

A perennial wisdom?

The perennial philosophy, the philosophical and spiritual heritage of all mankind. Of which Guénon is the greatest modern exponent. Guénon is mainly known as a Vedantist, an exponent of the Indian Vedanta. In the West, Western ideas, Plato is nearest to the Indian Vedanta. In fact, the core teachings of Vedanta are a traditional viewpoint, world-view, found across cultures and civilizations, implicit in religions, mythologies. It is important to reconsider—or rehabilitate—Plato in the West. The Islamic scholar S.H. Nasr said this. The most urgent intellectual task in the modern West, he said, was to rediscover the traditional wisdom, the *sophia perennis*, in the Greek philosophical tradition, specifically Parmenides, Pythagoras, Plato. Because this is where the West went astray. It turned against its own *wisdom tradition* and its own traditional cosmological foundations. It set out on the path of deviation. The dam burst. The West became disconnected from the rest of humanity. And suddenly, in Western ideas, the Greeks were presented as the precursors of modernity, the founders of modern science, the fathers of modern democracy. And Plato, after World War II, was demonized. Has been demonized throughout the second half of the twentieth century. Karl Popper, you know, cast him as the father of Nazism in an extraordinary misconstruction. It is essential to revisit Plato and to rediscover the heart of the Western metaphysical and cosmological heritage. There is no intellectual task so urgent. It is important that the West comes to a deep appreciation of the Islamic heritage—that is important too. That is an urgent intellectual task. But the rediscovery of Plato as the Western spokesman of the perennial philosophy, that is essential. And that is why I have studied Plato and use Plato.

Why not Pythagoras?

Parmenides and Pythagoras are important too. But Plato is the place to start. Plato, anyway, is a Pythagorean. And a Parmenidean. He gathers these traditions, strands of tradition, together. Weaves them together. We have only fragments of Parmenides' work, anyway, and it seems Pythagoras wrote nothing. Plato survives. And his works articulate a version, a rendering, of the *sophia perennis*. And, in particular, this is so in the Platonic myths. Athens is the place to start. The Acropolis is the place to start. We need to reclaim the Acropolis for Tradition. Instead of the fountainhead of science and liberal democracy and modernity, we need to reclaim ancient Athens, Plato's Athens, as a fountainhead of Tradition, with an upper-case T. Meaning a body of universal ideas. A common *global* heritage. For these global times. Found in India. And China. And Islam. And the ancient and medieval West. And in the primal traditions. Plato is central because he is the foremost spokesman of this tradition in the West. There is an entire dimension of Plato that needs to be rediscovered. Plato is the father of Western cosmology.

Not Aristotle? Surely the four elements of Aristotle—fire, air, water, earth—are basic to traditional sciences in the West, before the modern era?

But the four elements—yes they are basic—they are wrongly ascribed in Aristotle. They are—in our known sources—first articulated by Empedocles, and they become central in Western cosmology through Plato's *Timaeus*. Not Aristotle. In the Middle Ages there were summaries of Plato that were attributed to Aristotle. So Aristotle's name, his reputation was huge. But in modern times we have been able to distinguish between what was Aristotle's and what was Plato's and, frankly, the interesting part of Aristotle was Plato. The four elements of traditional cosmology are more Platonic than Aristotelean. Because Plato knew the secrets of Athens. The secrets of the Acropolis. In the central shrine of Athens—the Acropolis—sacred to Athena, goddess of wisdom, goddess of philosophy, a sacred place since neolithic times at least—that is the real center. The heart of the tradition. Plato was a loyal son of Athens and hidden in his works are insights into the secrets of the cultus of the city. There were four gods sacred to the Acropolis cultus. Hephaestus, the smith, the fire god. God of the furnace. And Athena, goddess of the air, clouds. A cloud goddess. A wind goddess. An air goddess. And Poseidon, god of the seas, of water. And Gaia, the Earth goddess. The four elements of Aristotle? They are the four elements of the Acropolis cultus. The four elements come into Western cosmology—mainly—through the Acropolis cultus, especially via Plato's cosmological writings such as the *Timaeus*, and the Platonic myths. Important, essential dimensions of Plato have been lost because we have divorced Plato from the religious mysteries of his day and especially the mysteries of the Acropolis. We have lost the context. We need to restore the context.

What context do you mean?

The way to recover or rehabilitate Plato and to cast a new understanding of his work is to restore the cultic allusions in his writings. The context of the Athenian mysteries. Where is the source of this idea of four elements—fire, air, water, earth? These are universal ideas, surely, but the most important place to look in the case of the Western tradition is the Acropolis. The four gods of the Athenian sanctuary. Hephaestus. Athena. Poseidon. Gaia. Fire, air, water, earth. We can't understand Plato except in relation to the Acropolis cultus. And then we must remember that the Acropolis cultus was transplanted from Egypt. And is very, very ancient. The key idea here that needs to be understood—and lifted out of sociological and political contexts and so universalized and understood as a spiritual concept, a spiritual notion—is the mystery of *autochthony.* Plato alludes to this all through his works but in modern times, modern readers, abstract Plato, turn him into an Oxford don. We read him as if he was an abstract philosopher. Whereas all the evidence is that, for example, he had a deep acquaintance with the crafts, connections with the soil. An earthy man. And a loyal Athenian acquainted with the rites of the Acropolis. And concerned to explore the inner meanings of those rites.

What is autochthony?

Autochthony is the great mystery of the Athenian religion. Literally it means 'native' or 'aboriginal'—born from the Earth as opposed to being an immigrant. This is the boast of the Athenians. They are *autochthons,* natives of the soil. But there are spiritual mysteries behind this boast, and Plato's myths illuminate these spiritual mysteries in the deepest possible terms. People spring up out of the Earth often in Greek mythology. Is it Jason—I think it is Jason—who plants dragon's teeth and they grow into armed men? But the important myth, the central myth... it is the central mythology of Athens, the Athens region. The Attica region. It is pre-Athenian, of course. It is very ancient and stems from an agriculture cult in Attica, the region around

Athens. Then it is overlaid a hundred times as Athens grows into a city and then an empire. But it is the original cultic mythologem of the Acropolis in Athens, and the Athenians adopted it as one of their key foundation myths. That is, the Athenians put it to political purposes. That is, it was State-endorsed.

Yes, surely it would be used to disadvantage immigrants? It would be used as a form of chauvinism just as native-born people in many parts of the world resent immigrants?

True. And it has been studied thus in many studies. Sociologically. Politically. It is one of the ideological planks of Athenian imperialism. And their chauvinism, as you say. Because it allows native born Athenians to regard immigrants as inferior. But Plato has a universal perspective. In Plato's time there was already a pan-hellenic culture and so a pan-hellenic perspective. And Plato wants to look further beyond that. He wants to draw out the universal meanings of the Athenian state cult. He clearly understands this mythology of Attica, this arcane agricultural mythology—the doctrine of the earthborn—as universal in its scope, a local mythology, a local version of a universal doctrine. In Plato—especially the Platonic myths—we find certain teachings that reveal aspects of the *inner* nature of this doctrine and the cult of the earthborn in Athens. Significances which are far beyond the parochial and chauvinistic applications of it in mundane Athenian politics. This is why, in the *Timaeus*, Plato has an outsider—Timaeus of Locri—come to Athens to illuminate the Athenian myths. And why he has Parmenides of Elea visiting Athens on the festival of the Panathenea—to sort Socrates out—in the dialogue called the *Parmenides*. In these dialogues outsiders come to Athens to draw out the deeper and wider significances of Athens' own mythology. Especially the mythology of the autochthons, the earthborn. What is most important is to realize its universal dimensions.

Which are what?

Two things, for example. Firstly, the idea of rebirth. Spiritual rebirth. And secondly, the motif of resurrection. These are two essential ideas in the world-religions. Both of these are part of, implicit in, the doctrine of the Earthborn. Which is, more fully, the doctrine of the Golden Race. The autochthons are golden souls. The golden soul is earthborn. This is the basic teaching of the Platonic myths. The Attic agricultural mythology Plato understands to be essentially the myth of the Golden Race. And so, more broadly, it is an alchemical mythology. It is a mythology about spiritual transformation, about the transmutation of metals, the salvation of souls. The god of the Acropolis, remember, is Hephaestus, the Olympian smith. This immediately signals an alchemical mythology because alchemy uses metallurgy as its metaphor. And, further, the Attic mythology is about birth, about a spiritual obstetrics, primarily the mystery of the virginity of Athena and her motherhood of the Athenians. The great temple of the sanctuary, remember, is the Parthenon signaling parthenogenesis. The mysteries of virgin birth. These are all essentially alchemical motifs. Because alchemy is—in whatever tradition—a combination of metallurgical symbols and obstetrics. The question is, how to be born of the Golden race? How to perpetuate the golden race? Just as alchemy is the quest—the spiritual quest—for gold.

Rebirth and resurrection?

Because the Earthborn state is one of spiritual realization. Rebirth is a central idea here. How can one be born from the Earth like the golden race? Like the golden souled autochthons? Of course, to do so you need to be born again. You have your mortal parents, your father and mother, your *human* connection. But that obscures your *spiritual paternity*. Spiritually you are not the children of your mother and father but rather the child of the Earth and Heavens. The Earthborn—this is the universal dimension—are those who *transcend* their physical birth, their mortal identity, and realize their *true* origins. In the Gospel

Christ said you must love God more than your mother and father. People find that a harsh saying of Jesus. But it is true. Your true origins are *beyond the parents of your flesh*. You can be reborn and when you are reborn you are born into the Earth-born state. This is the same as transmuting your leaden soul into a soul of gold. The reborn are the golden souls, the golden race. The perfected souls. Symbolically, you negate your own nativity, or rather *reconcile it with that of the world, the creation*. If successful, the Earth is then your Mother, the Sky is then your Father, the Air and Sea are your nurse. You transcend your physical identity. Becoming Earthborn means to be reborn, to be reborn as a child of the Earth. Its a rebirth experience. Or it is often described in those terms. The language of birth. Because it is about the mysteries of birth. Birth as a metaphor. Parentage as a metaphor. The Earthborn, for example, are described in Greek mythology as emerging directly from the Earth like plants. The first generation, the golden generation, were born from the Earth like plants. This symbolizes a birth *beyond the animal birth*. It is regarded as a superior mode of being born. It symbolizes a spiritual parentage. The realized Buddha is Earthborn. The enlightened Muslim is Earthborn. The Christian saint is Earth-born. They have all been born again into a new state. Which is actually their original and true state, not a new state at all. Different religions contain the doctrine of the Earthborn and express it in different ways, but it is the same doctrine in every case. The language of *rebirth* is religious language. There is obviously a religious dimension to this. But the religions are just different languages, different organizations of symbols. They point to the same thing. You find appropriate symbols in Islam, Christianity, pagan religions. But it is a universal experience.

Resurrection?

Many religions—Christianity, Islam, Judaism—contain the motif of resurrection. The symbolism of death accompanies the symbolism of rebirth, of course. To die and be born again. In these religions, at the end of time, at the end of the present cycle of time, the dead will rise from their graves, resurrected in the flesh. The same idea is found—more fully—in Plato's myths. In the myth in the *Statesman*, the *Politicus* myth. Time moves one way—forwards—according to the myth. But at the end of time it reverses and moves backwards. The Sun will rise in the west and set in the east. Then the young will grow old and the old... the dead will be born from their graves, the dead will rise from the Earth. Like plants. Their corpses are seeds. And at the end of the time they will be born again, afresh, regenerated, resurrected. Born from the soil. Earthborn. Autochthons. This is the new Golden Age. After the full degeneration of time, at the end of the Iron Age, is the End Time cataclysm, and then the new Golden Age. This is a very potent and ancient and widespread mythology, described by Mircea Eliade in his book 'The Myth of Eternal Return', for instance. Plato situates the mythology of Athens—Attica, the Attic agricultural mythology—within this wider mythos. This is really the key to understanding Plato.

Everyone is a native of the Earth. What do you mean 'Earthborn'?

Being Earthborn is a spiritual state, not your physical status. Certainly all human beings are children of the Earth... even if increasingly very few of them behave like it. Modern man is heavily abstracted. He lives in his head. He's lost that vital connection, that *belongingness* with the Earth. The idea of Mother Earth is just a platitude, not a modern reality in people's lives. What would industrial man know about Mother Earth? The earthborn are children of the Earth and behave as such. *It is a quality of being.* Citizenship might be automatic on Planet Earth but *aboriginality of spirit* must be cultivated. In alchemy—the alchemical mythos—the metals, the metallic ores, are embryonic in the Earth, in the womb of the Earth, and slowly, over geological time, these embryonic ores mature and change and ripen and develop, like embryos in the womb of woman. From the leaden state they slowly ripen into the golden state. And the art of alchemy is to take the raw ores from the womb of the Earth and to place them in an artificial womb, an *athanor*, and, by craft, to hasten the maturation. Because in the course of time all leaden souls will ripen again, but a spiritual alchemy is concerned to hasten this process. Mortal, fallen man has fallen into animal reproduction and he is born out of phase, so to speak. But in this lifetime—this lifetime is a great opportunity. That is what the great sages and traditions tell us. The blessing of this life is that in this lifetime we can make great process, fast. Like the alchemist maturing metals. In this lifetime we have the chance to die to the world and to be reborn in the spirit. This is the unanimous teaching of the perennial tradition. The Prophet Muhammad said, 'Die before you die!' What does this mean? It means—be reborn! Life gives you that chance. Transcend your mortal birth. Be reborn, *truly*. Make yourself right with the *kosmos* and with God, with creator and creation. Gnostic realization is always three-way. Creator. Creature. Creation.

I've never encountered the idea of 'Earthborn' before. Yet you present it as a universal idea?

Classical scholars—the humanist tradition—has been preoccupied with other aspects of the Greek heritage and has neglected this. But the ideas are not so unfamiliar. Only the terminology is unfamiliar. The idea of resurrection is common enough, certainly in a Christian context. And Jewish and Islamic. In the Abrahamic traditions. At the end of time the dead will rise from their graves, born anew from the Earth, *like plants*. But there is always the ambivalence between resurrection *of* the flesh and resurrection *in* the flesh. Let us explore this idea of resurrection. That will lead you into the mythology of the earthborn. In many forms. In many traditions. Hellenic and semitic. East and west. Hermetic. Alchemical. Once you see it, you realize it is a pervasive theme, a pervasive set of symbols and metaphors. It is a lost key. A key to a primordial alchemy. A key to ancient astrology. A key to religion.

How is it a key to ancient astrology?

Well, you need to keep in mind, always, that astrology is the sister science of alchemy, and in both cases we are trying to recover a lost heritage, a heritage that has become sullied with superstition, degenerated into superstition. But astrology can be understood as a language of spiritual symbols. There is a late ancient text by Julius Firmicus Maternus that preserves a system of very ancient astrology. We don't know much about Firmicus Maternus. A pagan. He became a Christian later in life. He wrote a couple of things including 'On the Theme of the World' where he describes a system of astrology which he says has been preserved in documents since ancient Egyptian times. He says the Egyptians knew the horoscope of the World. He preserves for us the horoscope of the world, or the Mundi Thema.

I haven't encountered that idea before either. A horoscope of the world?

The Theme of the World. The Mundi Thema. Theme is an old fashioned word for horoscope or nativity. In the Firmicus Maternus system of astrology there is a horoscope of the world. It is, as he says, a *symbolic device*, not to be understood literally, but he says it is the master key to all astrology. Modern astrology has lost contact with these ancient roots and so even if you are familiar with astrology it is unlikely you will know much about the roots of it as a traditional science or its place in the sophia perennis. It has become abstracted, unconnected, corrupted, decadent. The traditional planetary rulerships have been obliterated, for instance, yet they contain a surprising wisdom that modern astrologers don't even suspect. The idea of the horoscope of the world fell well and truly out of favour in the post-Darwinian era because, of course, it implies a creationism. The idea is that the world was created on such and such a day at such and such a time. Or it means that if you take it literally. Firmicus Maternus insists it is a 'symbolic device', he says. It represents the *pristine moment*. The moment of creation. The germ-moment of the world. In these post-Darwinian times it is necessary to stress that the idea of 'Day of Creation' in this philosophy is a symbolic one and should not be taken literally. The 'Day of Creation'—and the horoscope for the day of creation—is simply that hypothetical moment (a temporal symbol) when all the forces and forms of the world were pristine and static and held in an initial equilibrium. Properly understood this 'Day of Creation' is not some moment in the distant past, nor even the distant future, but is the indefinable, ungraspable moment *between* every moment in time. We are not remote from this Creation but rather are sustained by it every instant and always have access to it because it is ever-present; it is, at least, *implicit* in the world as we find it. The idea itself is an intuitive realization and not a statement of materialist science. It matters not at all if there was no 'real' Day of Creation such as many religious fundamentalists preach about. If you understand that, the idea of the horoscope of the world is not that odd. But its not surprising that modern people have never heard of it. Modern people think ancient people were stupid. This is a way of not having to think about it.

So, how does this horoscope of the world relate to the idea of the earth-born?

The system described by Firmicus Maternus is simple. Given the nativity or horoscope of the world, the astrologer must compare all individual horoscopes to it, and the prognosis and analysis of individual charts consists of comparing and contrasting it to the master key, the horoscope of the world, the Mundi Thema. A simple example: in the horoscope of the world the Moon is in Cancer, her domicile or natural house. If someone is born with Moon in Cancer they are in harmony with the pristine forces of Creation in that respect. If, however, they have Moon in Capricorn, 180 degrees from Cancer, they have a relationship of opposition to the primary arrangement. Their task, then, will be to reconcile this. They won't have a choice. Life will present those opportunities in which they can reconcile this. To fully reconcile your nativity to the cosmic nativity, the horoscope of the world, makes one *Earthborn*. You have transcended your individual identity, your individual horoscope, and are reconciled with the Earth. You have transcended the circumstances of your birth and been born again—direct from the Earth like a plant, Earthborn, as Greek mythology would have it. If you reconcile your individual nativity then suddenly you have the horoscope of the world as your nativity. That's the point. Everyone has two horoscopes, so to say. There's your physical horoscope. But then there's your *spiritual horoscope*. And all human beings have exactly the same spiritual horoscope, the same spiritual lineage, the horoscope of the world. Firmus Maternus describes it as being born 'with' the Earth. Greek mythology describes it as being born 'from' the earth. These are the same thing. Just as resurrection *of* the flesh is the same as resurrection *in* the flesh.

I don't quite follow. What do you mean two horoscopes?

In the center of the traditional horoscope is a square into which is written the name and birth date of the native. This square represents the ego, the false ego, of the soul whose horoscope is depicted. This is the animal birth. Ourselves as the product of our mother and father. It is name our mother and father gave to us. At the same time it represents who we think we are, our assumed identity, the name and person we imagine ourselves to be. Our true identity, however, is *hidden beneath this square*. The square obscures the true center of the horoscope. It is the task of every human being to go beyond their assumed identity, their animal paternity, and to discover their true identity, to shed this false ego and to discover their real being. This is the teaching of all the great spiritual and religious traditions throughout the world and throughout history and it is basic to astrological symbolism. These traditions all teach that human beings habitually invent a false image of themselves—we form a thought of ourselves and call it 'I'. But this is an illusion. This 'I' that we think of as ourselves is not real. It is a phantom, a mistake. It is just a mask that we wear. And this false 'I' hides our true identity, our true being, from us. And our true parentage. In a horoscope the false ego is the name written in the central square of the chart. This is the illusion that prevents us from discovering ourselves and discovering our real parents, the Earth and the Sky. Our true identity is hidden beneath this square in the center of the chart. Similarly, the whole horoscope shows the obstacles we face in the quest for self-understanding. We are our own obstacle. As the Muslim sage Hafiz said, *'You yourself are your own obstacle—rise above yourself.'* And Rumi said, *'If you could get rid of yourself just once the secret of secrets would open to you.'* This is not a doctrine of self-hatred. It is a doctrine of self-transcendence and of love. Self-negation, letting go of the false ego, is what every lover does. Ansari said, '*When you lose yourself, you find the Beloved. There is no other secret. I don't know any more than this.*' If a horoscope is at all useful it is as a diagnostic tool that shows us the shape and character of the false ego and that alerts us to the pitfalls that await us in the quest to overcome self-illusion. No one can avoid this quest. The entire human narrative is the quest for liberation from self. We might postpone the quest but we can never avoid it entirely—it is the human lot, the inherent conundrum of the human state.

How is this expressed in the horoscope?

One of the secrets of ancient astrology—almost entirely forgotten or corrupted now—is a certain configuration, a certain alignment of factors in a horoscope called the Pars Fortuna, the Part of Fortune. In ancient astrology it was considered the most essential thing in a nativity. But it was lost in Western astrology, only surviving in Arab astrology throughout the Middle Ages. Then it was brought back into the occidental world, Western esotericism, but in a corrupted form. Its real significance was forgotten. This configuration is the key to discovering what lays

behind the central square. In Western astrology now it is a mystery. No one is quite sure what it signifies, except that it has received the name 'Part of Fortune' meaning 'lucky' or 'fortunate'. Or 'blessings'. It is a position in a horoscope or nativity calculated from the positions of three factors, Sun, Moon and Ascendant. Or—to be exact—Sun, Moon and Earth, because the Ascendant here represents the Earth. So it is a synthesis of Sun, Moon and Earth. Even the ancient astrologers such as Ptolemy had only a superstitious understanding of this point, the Pars Fortuna, but in fact it is the *autochthony* point in a nativity. It is the key to spiritual rebirth. It is the point in a nativity that signifies the spiritual path, the point of transcending the animal birth. Sun, Moon,

Earth—this is the synthesis of the Earthborn. In any given horoscope the so-called Pars Fortuna is the real key.

I don't see how it is a matter of overcoming what you call the animal birth?

In the *Republic* Plato asks, Why does the perfect state fall into decline? Why does the Golden Age decline into the Silver Age and then the Bronze Age and then the Iron Age? And the answer, we are told—surprisingly, strangely—is unseasonable marriage. Because, one day it happens that children are born at the wrong time. Out of season. This is the so-called Nuptial Number in Plato, Plato's most famous mathematical conundrum. All of this mythology concerns the idea that man has fallen into animal reproduction, here symbolized as unseasonable reproduction. Our parents, they spawn our ego, our false self. But within us is still that golden seed, the golden soul, not the product of the flesh of man and woman and their unseasonable coupling but instead the product of the union of Earth and Sky, the autochthonous soul. Again, these matters take us into the mysteries of the Acropolis cultus. The rites of the Acropolis make a great play on Athena's virginity. The citizens of Athens, the autochthons, are really children of Athena, yet she remains virgin. She is virgin mother, just like the Blessed Virgin Mary in Christianity. The cultus of the Acropolis plays with this paradox, the virgin mother. But every noble born Athenian knows that—as well as their mortal parents—Athena is their mother. How? That is the mystery.

Is Athena the Greek precursor to the Virgin Mary in this regard?

Indeed, yes. These motifs are at play in Christianity too. With Joseph, the carpenter, the demiurgic figure displaced in a surrogacy myth of spiritual paternity. In Christian realization where it is 'Not I, but Christ in me!' Mary becomes our Virgin Mother. The mystery of parthenogenesis, understood—that is—as a spiritual concept. The true soul—as opposed to the false ego—has divine parentage. When we attain to the true soul, the true man, we transcend our mortal parents. And so behind every mortal coupling there is a hidden, a secret, marriage. Several ancient sources tell us that the Athenians regarded the herb violet (viola odorata) in high regard. One report says that large areas just out-side of Athens were devoted to its cultivation. It was also worn as a floral head-band by Athenians and featured in the rituals of the Acropolis. By all these accounts, it was regarded as a sacred flower, emblematic of the city and the goddess Athena. But what is special about the violet? Why did the Athenians regard the violet as sacred? Why was it used in the rites of the Acropolis and feature in the Athenian religion? Why was it an emblem of Athena? When we look into the biology of the violet we dis-cover that this herb hid a secret. It produces two sets of flowers, spring and autumn, but the two sets of flowers are different. In spring they are fully formed and sweet-smelling, yet barren. But in autumn the plant produces very tiny flowers, unscented, unseen, hidden beneath the leaves, with no petals, no scent, but super-fertile. Very few plants do this. It is a characteristic called 'cleistogamous', meaning 'hidden marriage'. This is the secret of the violet. This is why the violet is sacred to Athena and her rites on the Acropolis and why the plant was held sacred by the Athe-nians. This is why the Athenian horticulturalists grew acres of violets. It symbolizes the secret nuptial. The barren spring flow-ers represent the mortal parentage, barren because it produces the false ego. As opposed to the hidden flowers of autumn. It is the secret flowers, the hidden marriage, that is truly fertile.

How does the autochthony mythology come from Egypt?

Most obviously in the cult of Neith, the Egyptian equivalent to Athena. In the Nile delta, the city of Sais. That is the connection our sources make between Athens and Egypt. According to Herodotus. Neith is the Egyptian Athena, or rather Athena is the Greek Neith. And Neith was a weaving goddess—like Athena—and as such she wove the bandaging for mummies. She was inti-mately concerned with the processes of mummification. Mum-mification is like the process of *weaving a new skin*. And no doubt the process, at least in part, intimates the weaving of insects into *cocoons*. Underlying this is again the mystery of autochthony. In

Egypt we have a very sophisticated autochthony cult based on cocooning symbolism taken from autochthonous insects. The purpose of the cultus was to pursue the golden race. Note the use of gold in funerary masks in Egypt, for example. These arts involve making new bodies from metals—gold—and from weaving cloth. The mummies' cloth binding is the skin the Risen Man sheds. The gold outer casing of the complete mummy is the new resurrected body—the Golden Age man. Put simply, that is the basic symbolism. Then other things fall into place. Wallis Budge, the Egyptologist, for example, says somewhere... notes that the Egyptians buried their dead in the foetal position. For example. Exactly the symbolism of autochthony. Egypt is the hub of these very ancient teachings. Alchemical. They are transplanted into Greece. They re-emerge into Islam. They...

How do they re-emerge in Islam?

You can see it most obviously in the prayer ritual of Islam. The placing of the head on the earth. Rising up out of the earth, from the earth. But there are many motifs. There are many Athenian motifs that re-emerge in Islam. For example, in the Quranic

depictions of Paradise. The denizens of Paradise wear armbands of gold. Like the Athenians. Emblematic of their golden souls, their autochthony. And more generally in such a motif as the weaving of the peplos. In Athens, for example, at the Panathenea, the maidens of the city would weave a peplos, a garment, for the goddess Athena. This would be draped over a ship's mast and

then the ship, the Panathenaic ship, would be dragged up the slopes of the Acropolis as part of the rites. And every year a new peplos would be given over to the cult statue of Athena in the Parthenon. This re-emerges in Islam. Every year a new peplos is woven by maidens in Egypt—note, in Egypt, not Arabia—and carried to Mecca and it is draped over the Kaaba, the sacred shrine, replacing the old one. It is the same rite, but transformed, *mutatus mutandis*, for the Islamic symbolic vocabulary. Nevertheless, a very ancient weaving symbolism. The symbolism of warp and weft. It is the same symbolism as the symbolism of the ship.

Weaving and ship symbolism? How are they the same?

It is the symbolism of vertical and horizontal. That is the basic symbolism enacted in the Muslim prayer ritual. A movement between the horizontal plane and the vertical plane. Prostration and standing. Or, if you like, between sleep and waking. Christianity, of course, employs the same symbolism as the two axes of the cross. Which symbolism is made concrete in, say, the cathedral. But it is also obviously the symbolism of weaving where you have warp and weft, two axes. And in the ship, the axis of the mast, the vertical axis and the axis of the keel, the horizontal axis. This symbolism comes together very exactly in the rites of the Panathenaic ship in ancient Athens where the woven peplos is draped over the mast of the ship and the ship is dragged up the Acropolis to the sacred sanctuary. On a political level this ship represents the wealth and power of Athens. Athens as a trading power and a sea-going military power. But the ship also represents Athena and Poseidon working together, in unison. Athena represents the vertical axis. Sail. Wind. Air. Poseidon represents the horizontal axis. Keel. Sea. Water. It is the same symbolism—vertical/horizontal—as in weaving. And it has a deeper symbolism. A metaphysical symbolism that is the subject of Plato's *Parmenides* dialogue where, again, the myth of the earthborn becomes the key to understanding the work, in this case a very difficult work. In traditional cosmologies—which are an extrapolation from traditional metaphysics—the symbolism of warp and weft, the vertical axis and the horizontal axis, are central.

How is the myth of the earthborn the key to understanding Plato's dialogue Parmenides?

The *Parmenides* and the *Timaeus* are related dialogues. Both are set on the Panathenea—festival of the goddess Athena. Both are relevant to that festival and the gods and rites of that festival.

Plato is very careful to set his dialogues on appropriate occasions. In the case of these two dialogues the key to understanding them is the festival upon which they are set. The Panathenea is the great festival of the Acropolis cult. In Plato's time it had become a pan-Hellenic festival and people came from throughout the Greek world to witness it. This provides the dramatic pretext for Plato to have the aging Parmenides and Zeno in Athens where they meet young Socrates. Socrates then gives them his familiar yarn about the Platonic Forms and Parmenides proceeds to demolish it. This has caused scholars endless consternation over the centuries. It is agreed that the *Parmenides* is the most obscure and most difficult—but also most important—of all Plato's dialogues. But what can it mean? Parmenides seems to say that the theory of Forms—Platonic Forms—is wrong—it cannot be—but then he also seems to say that it must be right too—it cannot not be. What? Very puzzling. The key to the work is its setting on the Panathenea.

How? Why is it set on that festival?

Plato—very famously—has a grievance with Homer. Thus he is hostile to the poets in the *Republic*. But specifically, Plato feels that the appropriation of Homer by Athens—because Athens claims Homer and his epics and introduces readings of Homer to the sacred festival—has introduced an error into the Acropolis cultus, namely the error—from the *Odyssey*—that Athena and Poseidon are rivals. The error that the gods quarrel. So in this configuration of dialogues, set on the sacred festival—the *Parmenides* and the *Timaeus*—Plato teaches that the gods do not quarrel. This is the whole point of Solon's unwritten epic—the *Atlanticus*. The story of Atlantis is the epic poem that Solon would have written, should have written. Epic poems are recited on the Panathenea, so Critias tells of Solon's unwritten epic. The epic of Atlantis. It is a type of Athenian counter-*Odyssey* in which—most importantly—Athena and Poseidon are not rivals but complements. Plato won't have—cannot have—Athena and Poseidon as rivals. Homer does. In the *Odyssey* Athena and Poseidon are rivals, which is to say opposites. When

Homer is introduced into Athens this violates one of the core theological truths of the Acropolis cultus. So Plato offers an alternative to the *Odyssey*, namely the legend of Atlantis, Solon's epic. In the Atlantis legend the Athens/Atlantean war is not the result of divine rivalry but is caused by mortal error when the 'immortal blood' of the gods is diminished... and when autochthony is replaced with animal reproduction. But just as the *Timaeus* is a dialogue sacred to Athena, so the *Parmenides* is a dialogue of Poseidon. You have to read it very carefully and note the peculiar and unusual features which are the clues to the work. It is the strange, little unexplained details that give the clues. You need to read the beginning of the dialogue very, very carefully. And observe the Poseidonean motifs.

Such as what?

It seems irrelevant—a peculiar and unusual feature. But it is a key. Antiphon, after fully absorbing the Parmenidean doctrine, has retired to the horses, we are told. He has learnt all of Parmenides' doctrine and what does he do...? He goes off to tend his horses. Imagine that. Imagine mastering the metaphysics of Parmenides—the great father of Western metaphysics—and then, mastering it all, what do you do? Go and feed the horses. If the reader is not alert they miss this. They skip over it. Irrelevant detail. Padding. Filler. But it is key. Remember the setting. The Panathenea. As the dialogue is in progress, remember, the sacred rites of the Acropolis are being enacted. In such a context the reference to horses must surely allude to Poseidon, who rules over horses. A modern reader will easily miss this, but it would be far more obvious to an ancient Athenian. Panathenea. Horses. Poseidon. That chain of associations is very plain. And so too is the allusion to Hephaestus, the smith. Where is Antiphon? Antiphon is out instructing a smith in making a bit for a horse. If horse = Poseidon, smith = Hephaestus. And then at 137a in that dialogue Plato puts it very plainly. What does old Parmenides say? Parmenides says he feels like race horse and describes his task as 'traversing so vast and hazardous a sea.' Surely, the combination of these two metaphors—horse and sea—signals

Poseidon. God of horses. God of the sea. All of these—the dialogue is thick with carefully constructed allusions to Poseidon. We see them when we recall that the setting is the great festival of the Acropolis cultus. Plato only gives hints. These are sacred things. Mysteries.

How does this relate to the metaphysics of the dialogue?

Because the dialogue is essentially a metaphysical account of the dictum 'The gods don't quarrel' (or have rivals). Homer has allowed an unlawful opposition. A metaphysical error. When the Athenians—as part of their cultural imperialism— appropriated the Homeric epics, the nature of the Panathenea, the great festival, was changed. The Athenians introduced Homer into the festival. But Plato accuses Homer of a metaphysical error, of introducing an unlawful and heretical opposition into Athenian thought, in violation of the sacred teachings of the Acropolis cultus. Parmenides stands against Socrates' either/or, yes/no thinking. Parmenides' metaphysics is the 'true' meaning of the relationship between Athena and Poseidon. Athena and Poseidon are, in fact, mythological renderings of the metaphysical principles *essence* and *substance*. The vertical axis—metaphysically—is *essence*. The horizontal axis is *substance*. Socrates is right to say that essence and substance are contrasts. The world is not Heaven. Being and becoming can be contrasted. But they are not irreconcilable opposites. Beyond being contrasts, they are complements. Athena and Poseidon might be contrasted. Athena rules the ethereal realm. Poseidon rules the realm of flux. But *they are not enemies.*

So Socrates is wrong about the eternal Forms?

He is not wrong to make the contrast between universals and particulars, Forms and examples, but he is wrong to suppose that the these are antagonistic opposites. In alchemy, you must separate and purify your raw materials first, but then you blend them back together. Flesh and spirit are contrasts, but they're not opposites. This is a fatal flaw in externalist religion. To think that spirit and flesh are at war. It is true that as a first step in the spiritual process you must sharpen the contrast between spirit and flesh, and this might at first mean denying the flesh—but this is only a first step, not and end in itself. The spiritualist is as mistaken as the materialist. You sharpen the contrast in order to separate and purify your raw materials. *But you must reconcile them again.* And young Socrates doesn't understand this, which is why old, wise Parmenides says that Socrates has only just started out on the road to truly philosophical thinking. And why he says that Socrates is both right and wrong, much to the consternation of professors of philosophy ever since. There is no truth without paradox. Socrates wants to escape from paradox, but there is no truth without paradox.

How does this relate to autochthony?

Autochthony is what Plato otherwise calls the 'philosophical nature'. There are certain souls, he says in the *Republic*—the souls of gold—who have the philosophical nature. What is the philosophical nature? It is what is called—mythologically speaking, in the Atlantis story—the 'immortal blood' of the autochthons. Aristocracy—properly understood—is a prolongation of autochthony. All royal lines, you see, go back to an autochthon. And—here is the point—the philosophical nature is nothing more or less than *the innate capacity to reside in paradox... without falling into duality.* To completely internalize this, to make it your nature, to escape from false dualities, false oppositions, this is philosophical, spiritual realization. Then Heaven and Earth are aright, and you are in right relation to them, *a child of earth and starry heaven.* In most traditions this will be described as a reconciliation of the waking self with the sleeping self. Of night with day. The Sun with the Moon. Male with female. The reconciliation of opposites. Which is not the canceling out of opposites. The contrasts remain. The waking self is still in contrast to the sleeping self. Male is in contrast to female. But at a deeper level they are reconciled.

The traditional cosmological outlook is geocentric. How can you justify that in the light of modern science?

By making an appeal to a more complete view of things. The geocentric view of the *kosmos* is right, is true. And so is the heliocentric view. The Ptolemaic model is right as far as it goes. And so too the Copernican. And the Newtonian. And the Einsteinian. These are not exclusive to each other. They are different orders of truth. But the geocentric world-view is especially important because it is what offers itself to our direct vision, our direct experience. It is especially real in that way. We experience the world, the *kosmos*, geocentrically, with the earth at the center. It is what we see. It is what we experience. And we merely need to assume that there is purpose, or facility, in our experience. Whereas the modern models—true though they may be—are abstracted. Abstract models. You can't see them, or experience them, you can only think them.[1]

> *But it is worse, isn't it? The traditional outlook is not just geocentric but anthropocentric. It is assumes that man is at the center of the universe.*

Yes, and that man is the microcosmic reflection of the greater *kosmos*, of the macrocosm. That follows from the geocentric cosmology because it places man the observer at the center of things.

> *How can you accuse modern man of self-flattery when traditional man thought that he was the pinnacle of the whole creation? Now we know that man is just a lucky primate on a tiny, unimportant planet.*

These are ironies. Traditional man placed himself—or the human observer—actually the human consciousness—at the center of creation, as man the microcosm, and yet maintained a human-scale world-view and a human-scale technology. Industrial man belittles himself on the one hand by insisting that he is just an animal and that the earth is just a speck of dust in a massive universe, yet this is cover for truly Promethean self-

---

1. Wolfgang Smith deals exhaustively with the sense in which the geocentric theory is valid in his *The Wisdom of Ancient Cosmology: Contemporary Science in Light of Tradition.*

aggrandizement, conquering, raping nature. The traditional world-view is actually consciousness-centered. At the center of the universe is human consciousness. In the modern world-view it is ironic—a tragic irony—that the same science that relegates man to an animal on a speck of dust promotes him to a god who sees fit to build nuclear bombs and splice fish genes into strawberries according to his whim.

How can you explain the traditional world-view to modern man in the face of the scientific facts, and without recourse to religious explanations?

Once again Plato is useful here—if we revisit Plato and the Platonic cosmology. In the *Timaeus* he tells us that the two basic things in the *kosmos* are earth and fire, fire and earth. The element correlatives of Gaia and Hephaestus, gods of the Acropolis cult. But there is a very profound teaching hinted at—only hinted at— in those sections of the *Timaeus*. By fire and earth, the two primal elements, Plato means 'radiance' and 'solidity', as he says. Radiance and solidity are the two primal things. Other traditional cosmologies tell us the same thing. And to a modern audience we need to explain that here these traditions are hinting at a very profound truth, and this truth has to do with the nature of light. Because, by 'fire' these cosmologies are really taking about light. Smokeless fire, they might call it. They mean light. The great secrets of traditional cosmologies have to do with the secrets of light. And specifically, as here in Plato, they are aware of an inner paradox in light. We can put it this way, to suit modern ears—*as light slows down it shines upon itself.* This is the paradox. We can say— the whole universe is made of light. But light has two properties. Radiance and solidity. Within light there is a paradoxical nature. It shines *and it is that which it shines upon.* Light is invisible until it reflects upon a surface. Upon solid things. Yet these solid things—they are really made of light too. There is only one substance, light, and it has two modes—radiance and solidity. It shines *and it is that which it shines upon.* That is what Plato understands when he talks about fire and earth being the primal elements. In order for there to be a visible, tangible *kosmos* you must have radiance—light—and you must also have that which light shines upon—which is really light as well. Light shines upon light. This is a Quranic formula. The Qur'an states it very plainly in a very famous verse—*light upon light!* it says. This is one of the great keys to the spiritual teachings of the world. Light shines upon light. The stuff we call 'matter' is really a mode of light. Light has two modes. Radiance and solidity. Fire and earth. The two primal things. Scientific man can understand that the spiritual systems of the world, and the ancient cosmologies, are really about the nature of light. Modern man, Einsteinian man, understands that light is the key, the mystery, the mystery of the universe. Consciousness—the consciousness of man—is the universal light internalized, reflected inwards.

What do you mean? I don't understand.

Any consideration of 'consciousness' presupposes and alerts us to the subject/object distinction. We experience consciousness 'internally'. It appears to be an internal phenomenon and related to our own subjectivity. So this should give rise to the question: what in the macrocosm corresponds to this internal—and therefore microcosmic—faculty? What in the outer world corresponds to this inner faculty? The answer to this, surely, is light. We experience consciousness as an internal light. Light is the thing in the external world that provides the most adequate comparison with the internal faculty. Consciousness, then— whatever else we might say about it—is an internalization of light. And light is the best metaphor to apply to it. To formulate it: *Insofar as the human microcosm is an extraction or condensation of the human macrocosm (the kosmos) consciousness is an extraction or condensation (internalization) of light.* This is an important observation, an important principle. It is a key to understanding traditional pneumatology, or traditional psychology, the traditional understanding of consciousness. The light metaphor is paramount. And importantly, cosmologically speaking, the various subdivisions and distinctions we make within or about consciousness— 'ego' for example, or 'levels of consciousness'—will correspond to such categories as: Sun, Moon, stars, direct light, reflected light, shadow. And so on. This is a very extensive metaphor found in most traditions. The light metaphor—it can be extended to an array of other visual symbolisms. Mirror symbolism, for example. Consciousness, the mind, as a hall of mirrors. Also color symbolism. This is the way to understand such symbols as the Book of Revelation's '...*with a rainbow above his head*' (Rev. 10:1)—very much an image of consciousness as a mode of light.

Like the halo in Christian art?

Yes, also the convention of the 'halo', the luminous nimbus in sacred art. Not only in Christian art. Or in Islamic art, the Persian miniatures, where the Prophet is surrounded by a luminous flame. Again, fire equals light. And light is a metaphor of consciousness because in the human microcosm man has internalized light. The traditional way to understand this is as... The relationship between sentience and consciousness should be considered in the following way: A sentient being has an inside world—a microcosm—and an outside world—macrocosm—*and windows between these two worlds*. Windows. To introduce another visual/light symbol, namely 'windows'. Windows is another visual, light-based metaphor. One of the best discussions of consciousness is Al-Ghazzali's 'A Niche for Lights' which is a commentary on the 'Light Verse' of the Qur'an. The Islamic metaphor is light and veil. The veil metaphor refers to levels of consciousness. We are separated from the Light of lights by 70,000 veils, they say. Islamic sources here uses an implicit stellar symbolism. The 70,000 stars. Cosmologically—and paradoxically—the 'Light of lights' is *the darkness beyond the stars*—which are then the 70,000 veils.

How is the darkness beyond the stars the 'Light of lights'?

Well, it corresponds to the deepest layers of the mind. It corresponds to deepest sleep. Just as it corresponds to the black soil of the Earth. In many traditional cosmologies they imagine that the stars are pin-pricks of light shining from the Empyrean beyond, like pin-holes in a dome roof. Beyond is the Light of lights. In any case, in traditional sources, light is by far the most common metaphor. We certainly don't experience consciousness primarily as sound. In esoteric symbolism there is a third eye, not a third ear, in the center of the forehead. Light and visual symbols seems the more appropriate metaphor. Consciousness can be a silent gaze within. Note the visual language we tend to use in such discussions as these. Thus in this phrase 'It *appears* to be an internal phenomenon. . . .' 'Appears' is a visual word used to describe an event in consciousness. Thus also—less obvious—the word 'consider' which, etymologically, means 'to refer to the stars . . .' with the analogy 'stars equals thoughts' implicit. There is also, of course, the *'imagination'* which is explicitly 'the faculty of images'. Conventional science thinks of the human being as having 'evolved' from his 'environment'. Traditionally man is 'extracted' or 'condensed' from the *'kosmos'*—not 'evolved' from his 'environment'—the *'kosmos'* in this case being understood as the 'cosmic icon', itself a symbol and an 'exteriorization' of the Principle or as *a 'projection' of the contents of light*. Manifestation is a projection of the contents of light. And the contents of light—the inner nature of light—is paradoxical. Because light is both radiance and solidity. Fire and earth. Radiance and reflection. Outside and inside. A mystery.

> *But why should the earth be regarded as special? Science knows it is just a speck of dust, an insignificant satellite of the sun.*

The answer to that question has exactly to do these same matters—to do with light. Because the most striking feature of light in the human macrocosm, in the *kosmos*, man's environment—the geocentric world—is *the proportion of reflected light to direct light*. Or moon to sun, sun to moon. In a relationship that is, as far as we know, unique in its freakish improbability. The strangest of coincidences. As far as we know, this is—in that respect—the weirdest place in the whole universe. The Sun/ Moon polarity is the answer to this question. And then we remember that one of the first facts of human consciousness is that we experience it in two—at least two—main modes: waking and sleeping. Because dreaming is a mode of consciousness. This is the microcosmic parallel to the Sun/Moon polarity. Why do we need to sleep? Why is consciousness bifurcated like that? Why do we have cycles of sleeping and waking? Why that primal duality? Why is that duality built-in at such a fundamental level? In short, *because we are a product of a dualistic environment.* A dualistic environment of light. Night and day. But—equally important—direct light, reflected light. Sun and Moon. The question is 'Why is this solar system special?' and, more especially, 'If man is the ontological axis of the universe, why is his home situated at some remote corner of an insignificant galaxy?' The first step in answering this question is to clarify the traditional position. The traditional assertion is that the Earth and Man are central in Creation because *both are paradigmatic.* That is, they express something *quintessential about the universe as a whole.* What might that be? Why is this particular solar system special? Because here something of an essential and paradigmatic nature takes place that is central in the totality of things though seemingly peripheral and insignificant on a physical, which is to say, cosmographical level. Scientific man, the industrial order, doesn't even have a *cosmology* as such—modern man merely has an extended, inexplicable cosmogeography. Which is really baffling for him. And in this vast cosmogeography—which makes no sense at all—he can make no sense of his own existence, least of all the very faculty of consciousness that he uses to explore this vast cosmogeography.

What do you mean by cosmogeography?

A science with no sense of proportion. Goethe said, very truly, 'The microscope will make us blind!' This is true. We have extended our view inwards with the microscope and outwards with the telescope. But at the same time we have lost all sense of a human-scale universe. The secrets of the *kosmos* aren't to be found in a telescope or a microscope—we seem to think that if we could just build bigger ones at greater magnifications all would become clear to us—but instead the truth, *the God-given truth,* is right there to the naked eye. On a human scale. At a human proportion. Phenomenologically. Whereas we have extended our view far beyond any meaningful dimension. The ancient *kosmos* is as it appears to the human eye, unabstracted. We have succumbed to an abstracted universe that makes no *human* sense.

What do you mean by abstraction?

Again, beyond the immediate human scale of things. In cyclic decline—the decline of the world, the descent of the world into quantity, quantification—there is a corresponding movement of abstraction.

Can you give an example?

When the modern man looks at a beautiful twinkling star and says to himself, 'That's a third magnitude white dwarf 42 million light years from earth....' Quantities. That's abstraction. Or, for example, a simple illustration of abstraction, the way even the sacred tradition has become abstracted, is to think of where in the body we locate the seat of the soul. Where is the soul? In what organ? Where do we *center* ourselves, in the physical body? Originally—at first, in the earliest layers—it is in the *liver*. The liver, which moves to a 24 hour, night/day cycle. The liver is the organ of duality. It is the day/night cycle internalized in the microcosm. The primal duality is internalized in the liver. Primitive traditions still retain this idea. And in the ancient world. But then—later—this physical seat of the soul is understood... it is moved to the heart. The heart becomes the center. The heart with its diastolic cycle. The heart where the night/day, nocturnal/diurnal cycle of the liver is contracted into the on/off heart-beat cycle. We can see this is Catholic iconography. In earlier iconography—Christian—Christ is pierced in the liver— hence the water and the blood, the two fluids. But in later Catholic iconography this is forgotten and it becomes his heart. It becomes the sacred heart—but in a sentimental form. And then, still later, the center of man is understood as *the brain*. Now the brain replaces the heart. So, in New Age spirituality it is the brain, the brain is conceived in some way as the spiritual organ. Liver first. Then heart. Then brain. These are increasing levels of abstraction. Each organ is less *physical* than the previous. The brain—no moving parts. Our self-understanding has become more abstracted. We start by counting on our fingers. Then we count on an abacus. Then, finally, on a calculator. The microscope makes us blind...

You were talking about what is quintessential to man and the earth...

Yes. Because it is important to understand, in the first instance, that the geocentrism and—if you like—anthropocentrism of traditional cosmologies asserts that, even if the universe is wide and broad, what occurs here on Earth, especially in the life of Man, is of universal and absolute and not just local and relative significance. That is, life on Earth *epitomizes the whole*. The traditional doctrine does not arise out of parochialism but is an assertion of a positive doctrine: Man and Earth are central because they are *a paradigmatic expression of the principles of the whole Creation*. This still explains nothing to the scientist, though, because to him it would seem that the only reason for thinking ourselves paradigmatic is self-interest and the only reason for thinking the Earth a place of paradigms is because we happen to be here. Doubtless, they say—so the argument goes—beings on other star systems share the same delusions about themselves and their home too. So the challenge is to demonstrate that the Earth is indeed *paradigmatic* and of cosmic and not just local significance. That it is truly a center. The center. Traditional cosmologies rarely provide such demonstrations because the paradigmatic character of the Earth is an unshakable axiom—it is axiomatic—and it is only in the decadent stages of such traditions that challenges arise. But a simple demonstration is implicit—everywhere implicit—in traditional sources and you can extract it and present it in modernistic terms if you like. Almost every word uttered by the sages of the world—the prophets, seers, philosophers, wise-men, saints—throughout the whole of human history strikes the industrial ear—this is the arrogant posture of modernity—as 'mumbo-jumbo', so let's refrain from using authoritative sources and speak in plain terms, prosaic terms instead.

So what is paradigmatic about our world?

Well, we could here make a direct appeal to the very fact that *such a question is asked* and then consider the nature of a consciousness that is moved to ask such questions. It is consciousness and not matter that is by far the most curious phenomenon in the universe and it is in meditation upon the nature of consciousness that the keys to deeper insight lie. But let us follow the scientist and insist upon starting with material forces and factors and proceed under the assumption—which turns out to be spectacularly wrong—that consciousness is an outcome rather than primary. So, the answer the scientist-type needs to hear is: *its environment of light.* There is in this world—this remote corner of the universe, whether by chance or design is not an issue at this point—*an environment of light* that is paradigmatic of the nature of light itself, and since light is the very stuff of the universe therefore also paradigmatic of the universe as a whole. Traditional cosmologies share a primary preoccupation with light. In religious metaphors. Most traditional cosmologies take it as a starting point. Most familiar is the *fiat lux* (Let there be light!) in the Biblical tradition, but other traditions hold the same. Sometimes light is described as fire and sometimes—as in the Hindu tradition—it is interchangeable with sound, with the Sun conceived as 'chanting' it's life-sustaining rays, for example. And sometimes traditions describe various grades of light, such as the 'limitless light' of the Jewish Qabbalist. But light is the agreed starting point. And it is light, as I said earlier, that holds the key. For the *environment of light* in our world is very peculiar, if not unique. This is consonant with scientistic data. In their quest for extra-terrestrial life the scientists chase after water molecules or evidence of oxygen on distant worlds believing that these are the conditions found on Earth that have permitted such a development as Man. But the scientists own data should tell them that these crude building blocks of organic structures are meaningless without a certain *context,* and to date such a context is only found in the terrestrial environment, and that it is the *extraordinary conditions of light in the terrestrial environment* that are the determining factor.

*Can you be more specific? What is peculiar about the environment of
light?*

I mean the extraordinary relationship that exists between the
Sun, the Earth and the Moon. This relationship is such that, as
either a solar or lunar eclipse demonstrates, the Sun and the
Moon are more or less the same dimensions from the vantage
point of the Earth—from the human viewpoint. It is really a fantastically improbable alignment. Because of this alignment the
Earth has, in effect, two Suns, the Sun itself and the 'Sun of
night', the Moon which, though casting no light of its own,
forms an almost perfect—certainly adequate—mirror of the Sun
from the terrestrial viewpoint. Thus the terrestrial environment
is bathed in two lights: the *direct radiance* of the Sun and the indirect *reflected light* of the Moon. All planetary environments will
receive some reflected light from their satellites, where they
have them, but the planet Earth is extraordinary in having a satellite that is so large in relation to its parent that it appears as
altogether sun-like to the terrestrial observer. No other two
bodies known in the universe have such a relationship. Such
proportions. This is what is peculiar about our environment of
light. To any observer from beyond our solar system it is such a
striking relationship that they would not hesitate to nominate
the Earth not as a single planet but as a dual planet, a double
planet, the Earth and Moon operating as a single system. Quite
apart from the improbability of such a dual planetary system
with bodies having such relative proportions to each other,
there is then the Moon's relative proportion to the Sun and
Earth, and it is this that gives the terrestrial environment its second Sun, a satellite that mimics the Sun, so to speak, duplicating
the Sun's apparent size and its function of illuminating the terrestrial environment. The terrestrial environment today is lit by
two sources: the Sun and the Moon, and—this is the important
thing—by *two types of light*, radiant and reflected. And these two
types of light are found in a particularly felicitous proportion
such that the Full Moon is a near perfect match for the Sun. So
you see the importance of this...

Not really. Can you explain?

The importance is: here in this environment, the sub-lunary realm, terrestrial realm, the proportions of light are such that there is an archetypal ratio, just the right ratio, of both types of light. Radiant and reflected. And thus the two natures of light itself. Traditional cosmologies propose—not unreasonably—that Light is the creative *materia*. Let there be Light! All the traditions say it. And what we call *matter* is a condensation of light. The crude matter we touch and feel is, the physicist tells us, far less solid than it seems, for it is just a particular density of atoms, themselves suspensions of energy. Light bounds matter. As matter approaches the speed of light its mass increases 'to infinity'—the flash of the first atomic bomb in the New Mexico desert—when a 'new sun' was born—demon-strated this surmise of Einstein's with chilling irrefutability. Then, next, traditional cosmologies propose that the Sun-Moon-Earth alignment exposes—*reveals* would be the right word—something fundamental, quintessential, about the nature of light. The 'doctrine' goes: *There is in this world an environment of light that is paradigmatic of the nature of light itself and since light is the very stuff of the universe therefore also paradigmatic of the universe as a whole.*

So the Sun/Moon polarity reveals what you called the inner paradox of light?

Yes. So traditional cosmologies draw attention to the paradoxical nature of this 'light/fire', for if what we call 'matter' is, finally, a condensation of light, we must confront the paradox that light—invisible in itself—may *shine upon itself* in certain modes. The Qur'an, as I said, presents this idea in the celebrated mystical phrase 'Light upon light'—the paradox of light shining upon light—an important inner theme in Islamic spiritual contemplation. And Plato has the same idea. Radiance and solidity are expressions of the same principle. There is *only light* but the nature of light is such that it 'condenses' into solid matter and *shines upon itself,* bringing forth *a visible manifestation.* There is radiance and there is reflection. *The mystery is that what radiates is also what reflects.* Really. The mystery is that what radiates is also what reflects!

And the Sun/Moon system illustrates this idea?

Indeed. This paradoxical characteristic of light—the primal stuff—is expressed—remarkably—in the Sun-Moon-Earth alignment. Observers from the surface of the Earth are in a perfect position to understand this characteristic, for their environment is bathed in the light of the Sun *and the light of the Moon*, radiance and reflection, and these *in their primal proportions.* So, this is why the Earth is cosmically central. It is cosmically paradigmatic, and specifically the terrestrial environment of light is paradigmatic of the nature of light itself, and in revealing this characteristic of light this environment reveals the nature of the primal principle. More exactly, it is *the symmetry of direct (solar) and reflected (lunar) light* upon the Earth, that particularizes the Earth. Again, light has, from the outset, two properties, 'radiance' and 'reflection'. The threefold equilibrium of Sun, Moon and Earth allows for these in *an exact proportion in the terrestrial environment.* Only a Sun (radiance) and a Moon (reflection) of equal size—as seen from the Earth—allows for this exact proportion—*it is the proportion that allows the maximum contrast between these two properties of light.* What is special about the Earth is that it is the point in the universe where, by the throw of the dice if you insist, the inherent polarity of light—the paradox of light—is given its most concrete expression. A total eclipse of Sun or Moon points to the Earth as if to say 'Here is where the polar nature of light, the inner paradox of light, will be demonstrated!' And so here on Earth—this is where the inner nature of light is emptied out, put on display. Where the very principle of Creation is made manifest. If light didn't shine on light but just shone, there would be nothing but a plenum of light with its possibilities still intact and unrealized. Instead, the possibilities are realized here on Earth.

And consciousness is one of those possibilities?

This is the source of consciousness. Consciousness is implicit from the outset in the polarity Subject/Object, light upon light. It is as if conscious life becomes *necessary*, because this extraordinary coincidence has to be witnessed. It has to be witnessed because it is inherently self-reflexive. Not all extraordinary coincidences in the universe need to be witnessed, but the requirement that this coincidence be witnessed is inherent. The essence of consciousness is *reflection*, or rather the 'equality' of Subject and Object, Sun and Moon, witness and that which is witnessed. The two species of light, direct and reflected, in their brief and improbable equilibrium, create on Earth an arena of polarities and contraries. The terrestrial *kosmos* is the world that expresses the *polar possibilities of light*. Reflection is also the essence of life. The essential biological operation of life is invagination. This is because light shines upon light and because God is 'I am that I am'. Life on Earth is inherently self-reflective.

What is invagination?

Many people try to marry traditional cosmology with the physical sciences—physics. Quantum physics, and so on. Wolfgang Smith does this.[2] We've been talking about light. But really it is the biological sciences that overlap most with traditional cosmology. Invagination is really the most mysterious process. It is that strange point, the point at which the expansion of living cells ceases to expand outwards and instead 'invaginates' which is to say turns inwards and *creates an interior space*. The key step in biological evolution is invagination, the turning back upon itself of the merely chemical life processes. Cells start to divide and multiply. But suddenly they circle back. They create an internal realm. An internal world. A microcosm. Cancer cells don't do this. They just replicate in a formless way. Cancer is a failure of the microcosm. A microcosm is an internal space, distinct from the outer world. And invaginated lifeforms arise from the environment provided by the threefold equilibrium of Sun, Moon and Earth. You will not find invaginated lifeforms anywhere in

2. See his *The Quantum Enigma: Finding the Hidden Key*, 3rd edition, Sophia Perennis, 2005

the universe where there is not *the symmetry of direct (solar) and reflected (lunar) light.* Or, put simply, you will not find life anywhere in the universe—earth-like life—except where you have reflected light like that of our Moon. Without our Moon there is nothing. Without that principle of reflection there is no internal life.

So there is no life on Mars?

Maybe. But not earth-like life. Although, one has to observe that it is altogether startling that there has been no life found outside of the terrestrial environment. On Earth, life seems to be ubiquitous. You find life adapted to every nook and cranny. There are even bacteria that live inside volcanoes. And sulfur-loving bacteria that thrive at the very deepest depths of the ocean. Everywhere. Life is everywhere. Teeming. Swarming. Irrepressible. But on the Moon, nothing. In space, nothing. On Mars, nothing. On comets, nothing. Not a sign of it. The terrestrial example would lead us to believe that it is everywhere. In every crevice. Wherever it can get a foothold. But not so, it seems. It is really surprising that life has not been found elsewhere, no matter how hard we look. Not a single instance of any life whatsoever outside of the terrestrial environment, the sub-lunary realm. It seems that the essential move of living things— invagination—the creation of internal space—is a product of the terrestrial environment and specially the terrestrial environment of light. Moreover—an important point—the invagination process in biology is the exact correlative of consciousness which is the *turning back upon itself* of the physical processes that are the station of thought. You will not find consciousness anywhere in the universe where there is not *the symmetry of direct (solar) and reflected (lunar) light.* Both consciousness and biological invagination are an expression of the same environment of contraries and reflections. Everything on Earth is an expression of the same environment of contraries and reflections. So, the things of the Earth—and especially the conscious being Man, who has internalized light—are of universal and not just local significance because they give form to, and play out, some of the fundamental possibilities inherent in the very stuff of the entire universe. The threefold equilibrium—Sun, Moon, Earth—is imprinted on the very fabric of life where it plays out all the consequences of *the symmetry of direct (solar) and reflected (lunar) light*. This arrangement of equilibrium itself invaginates, generating a microcosmic order that reflects the macrocosmic arrangement. Man and the human form is the fullest expression of this, the fullest invagination of the cosmic order itself. All the parts of Man are invaginations of the *kosmos*. The most *macrocosmic* expression of life is the plant. It has no inner life. Its organs are the Sun and Moon and stars, external to itself. The threefold equilibrium of Sun-Moon and Earth is found expressed in the flower-leaf-root organization of the plant kingdom. The Sun builds its vertical structures and the Earth its horizontal structures. The Moon governs its cycles and rhythms. The most *microcosmic* expression of life is Man. Man and plant, not Man and animal, are opposites. In Man we find the organization of the plant upside down and inside out. The processes of 'enlightenment'—the spiritual processes—involve turning *our given consciousness upside down and inside out*. These processes are not subjectively or culturally determined but are written into the very creation of Man.

Is alchemy independent of religion?

No, because... due to cyclic decline and the growing crisis of historical man the alchemical traditions wherever they were and are found still need to be dependent upon an integral revelation, and so alchemy is now dependent upon religion, the great religious orders. You can't be just an 'alchemist', say. The whole order of the world has reached such an impasse that that would be impossible. You can be a Christian alchemist or a Muslim alchemist, and so on. The great religious orders have woven together a matrix in which these things, the traditional sciences, can survive. But not on their own. The scattered pieces of broken traditions—ancient traditions—have been stitched together into the world-religions. That is the only place where these things survive. But even then—if I might say so—the religions are, as it were, necessary evils, emergency measures, symptomatic of the Iron Age.

What is the best way for people to rediscover the traditional cosmology?

For most people, through the crafts. And of the crafts, probably best through farming. We might say through prayer and religion but in fact there is no prayer so humbling as tilling the earth and working the soil. These things should be discussed through something concrete and real like farming. The human relation to the soil and to the plant. In the *Timaeus* Plato says, for instance, that Man is an upside down plant. He uses that image. It is found in Plato and in Hindu sources, and Jewish sources, and elsewhere, across traditions. What might this mean? It is a very profound idea and one of the great keys to traditional cosmological thought. But what does it might? You might come to conceptual understanding of this notion, but finally it is something you need to experience. You need to put aside all your sophistications of modernity, those abstractions, and grasp the simple reality of this analogy—Man is an upside down plant. Upside down. Inside out. Man—in his whole organization—is upside down and inside out in relation to the plant. How strange this seems to modern ears! What? What a weird notion! Like every other aspect of traditional cosmology it is something you need to experience. Look at the stars. Forget what science tells you. The sky is still a map and a book. Somehow you need to distance yourself from modern abstractions and rediscover the Real. Religion might help. At least it provides a framework. But the traditional sciences involve no faith. They are based in direct, primal human experience. There are two symbols of alchemy—the black soil and the darkness beyond the stars. Whatever it is, it is useful—always—to come at philosophical and religious things via some concrete problem. The created

order. The *kosmos*. This leads us to philosophical and religious matters. It is always wrong to start with metaphysics. That is not the place to start.

> *But our main interests here are philosophy and religion? And metaphysics is primary, isn't it?*

If you like, but the main interest is really to engage with life, foremost. Deeply and not superficially. And that *means* being philosophical and religious because you run into those things—philosophy and religion—when you engage with life. All you can really do is try to engage with life more fully and deeply. And escape from abstractions. Life prompts us to real questions. It is important to avoid compartmentalizing philosophy and religion as this very abstracted, very pretentious activity in which various experts engage. In fact, life itself will make you philosophical or religious or spiritual in a genuine sense. Many people don't engage much with life. It never makes them wonder. Whereas philosophy must begin in wonder, as Plato said. With *thauma*. Wonderment comes from engagement with life. Not wallowing in life but rather seeing the signals, the signs, and following the threads—having eyes to see. And if you do that you will find that the fundamental things in life are very philosophical and spiritual things. All the mysteries of life, the *kosmos*, are right there in the seed and the germinating seed and the plant and the beast and the sky and the soil and the wonderful alchemy of the compost heap and the rain. Even the compost and the manure and the earthworms. If you go into anything deep enough you run into philosophical and spiritual matters. In the Bronze Age somebody went into the mysteries of metal-craft and revealed that it was a deeply spiritual thing, and thus we have *alchemy*—which is a spirituality of metallurgy. Thus too there is a farmer's alchemy. And other craft-based philosophies and spiritualities. It is always best to anchor these things in concrete issues. We live in very abstract times. Wisdom comes out of the Earth, out of the soil, first of all.

Not from Heaven?

From Heaven, yes. But it is—at least—*mediated* through the Earth and the soil. And that is the important thing to realize and exactly the thing that modern man has forgotten. At dusk, when the earth breathes, when the smell of the tilled earth is sweet and damp in the air, and peasants imbibe it, draw it into their lungs—that's how wisdom enters into man. It's the exhalation of worms and the buzz of saw-flies. It's the breath of the living earth. It is *that* real. It might be traded around in schools and universities and written about in books and prostituted at conferences and talked about by sages, but it comes from, or *through*, the soil. Through the clay. Through the rocks. It is *that* basic, *that* elemental. An abstracted man has no part in it. Certainly modern man has lost his connection to Heaven, but *he has lost his connection to the Earth as well*. And the way back to Heaven is through the Earth, not the other way around.

> *But the modern crisis is a spiritual crisis, isn't it? The ecological crisis and other problems—these are symptomatic of a greater problem which is spiritual, isn't that so?*

It is, but you can't just leap up to Heaven in a single bound. Man didn't just lose his connection with Heaven. More specifically, *he forgot that his connection with Heaven was through the Earth*. That is the nature of the modern spiritual crisis. That is the whole problem. Man didn't forget Heaven. He forgot that, by nature, his approach to Heaven is through the Earth. Modern man is spiritual enough—or *too* spiritual. But its a raw spirituality of unmediated access to Heaven, as if the Earth and the soil is not itself spiritual. The way to metaphysics is through cosmology. This is what needs to be put back into perspective. Think, for example, of the Islamic prayer ritual. What is the way to God? *By pressing your forehead to the Earth*. The Muslim prayers illustrate this principle exactly, beautifully. In Islam you find God *by pressing your forehead to the Earth*. You don't raise your eyes to the sky and call God down. You don't try to leap up into Heaven. You put your forehead to the Earth because God comes to us *from the earth*, from the soil, from concrete, real things. It's an important para-dox. If a modern man wants to find spirituality, wants to find God, what is he to do? The best advice to give him would be to start digging a garden. Go plant a tree. Get his hands dirty. Touch something real. Sit out under the stars. Put away the tele-scope and just watch the Book of Creation. Contemplate his navel? No. Dig a garden. Or learn a craft. Take up carpentry. Woodturning. Metalcraft. Needlecraft. Anything like that. Gar-dening, or a traditional craft. That is *much* better advice. The search for God must be grounded. The vast majority of men and women are only suited to *karma yoga,* you know. The path of *work.* The path of the crafts. That is the heart of the modern crisis.

But modern man is too materialistic, too caught up in the things of the world. Isn't that true?

It's wrong to imagine that modern man is too materialistic, as if he is too *concrete,* too *real.* For all his *things* he has never been so disconnected from the Earth. It is wrong to say—modern man is too materialistic, therefore he should turn away from the physi-cal, shun the physical, rediscover the metaphysical. That is a fun-damental misdiagnosis. If a man complains that he is lost and cannot find God, after talking to him a while you'll usually find that this is all abstract talk. His real problem is *work.* He has no real connection to the Earth *or to Heaven* because he has no work to which he can devote his heart and soul and hands. For all his so-called materialism modern man doesn't feel at home on the Earth. *Man has never been so alienated.* Wisdom might come from Heaven, but Man has to find it in the Earth, from the Earth, from the soil, from real things. The only path to God for most people, the only path for which they are suited is *karma yoga,* the yoga of work, but the modern order, the industrial or capitalist order, keeps them from any proper work. They might be immersed in consumer goods, but these are actually abstractions. Modern man doesn't buy a car, he buys a *dream.* The problem is not mate-rialism. On the contrary, the problem is abstraction. It arises out of the capitalist modes of production. Men and women are *divorced* from the Earth. Wisdom, spirituality, comes from real things.

> *Then it is a wisdom of peasants and peasant farmers and labourers you are talking about?*

It is. You might think that alchemy, the alchemical tradition, is the preserve of bearded old wizards or strange secluded scholars toiling away in secret laboratories, or mad scientists—those stereotypes of the alchemist—but in fact alchemy is nothing but the *wisdom of blacksmiths*. At root. It is the spiritual wisdom of blacksmiths. Big, burly blacksmiths with huge hands, thick brows and slow, sedate eyes. The scholars might trade in the alchemical arts but those arts *belong* to the blacksmith. And there are other alchemies that belong to other craftsmen. That is what has been lost. There is an alchemy of farmers. It is important not to lose sight of this. The scholars are middle-men, or worse. Wisdom comes from the soil, or wisdom comes from the blacksmith's furnace. From the farmer's plough or the blacksmith's anvil. Not from the scholar's pen. You need to appreciate the paradox here. In the village it is *the most physical man*, the blacksmith, who is in possession of the great spiritual secrets. We live in such abstracted times it is important to reiterate this now. In alchemy, you know, the key to the transmutation is *lead*, the most base metal, the most physical thing. It is true etymologically. Consider the word *sophia*, wisdom. It means—its primary meaning—is *skill*. The craftsman's skill. In Greek mythology Athena, goddess of wisdom—she is goddess of weavers first, of craftsmen and craftswomen. If we are going to discuss philosophy and religion let's acknowledge these roots in basic things, crafts, first. Through cosmology. We don't even begin with God. God is *far* too abstract. The way to metaphysics is through cosmology. The way to wisdom is through the crafts. Modernity is a crisis provoked by the industrial and scientific revolutions which have stripped Man of any meaningful cosmology or any meaningful cosmic context in which to place himself. Modern man is disembodied. So, in some ways he is *too* spiritual, *too* abstracted. He lacks the Real.

But the 'Real' is a metaphysical idea. The world is just 'illusion' or 'maya', isn't that so? And metaphysics is the concern of religious traditions, not farmers?

The transcendent is Real. God is Real. And in *contrast*—if we need make the contrast—the world is illusion, *maya*. But only from one viewpoint. The contrast is not absolute. And—as a partial fact—it is not an invitation to a world-hating other-worldliness. To fall into that *dualism*—Heaven versus the world—spirit versus matter—is a simplistic perversion. The oldest of heresies. Although we might sharpen contrasts for certain purposes, as a step, just one step in a complete process, we should never de-physicalize the spiritual. The resurrection is in the flesh. A physical thing. We must avoid spiritism. We have *de-sacralized* the physical world in the mistaken belief that physical things, the world, is not spiritual. As if 'physical' and 'spiritual' are opposites—rather than complements. The spiritual is not *anti-physical*. Not finally. One of the most beautiful, subtle moments in Plato is when we hear that the man who has realized the whole of the metaphysics of Parmenides is out with the smith tending to the horses. It is a motif in Sufi stories too. The seeker arrives after a long journey looking for the Master. He goes up to the man who is quietly sweeping the floor in the forecourt of the mosque and says 'Can you tell me where I can find the Master?' But, of course, this man who he thought was the cleaner is the Master. He expects the Master to be some exalted spiritual being perched on a peacock throne deep in meditation. The Master is the man who sweeps the floors. Physical is not the *opposite* to spiritual. And so the integrity of a traditional civilization is, on one level, as dependent upon the maintenance of a good heritage of seed for the annual grain crops, and upon the good tilth of the land, as it is upon the maintenance of a legitimate transmission of religious mysteries. The priests and hierophants are guardians of the way to salvation, certainly, but a farmer or gardener is guardian of the seed and the soil where, in one sense, the *deepest of all mysteries* lie. The soil itself—the black soil—is a symbol of the bed of tradition; a living thing, structured, fertile; it supports the flowering of generation after generation while remaining unchanged. That is *tradition*. Tradition is the soil. Alchemy, the word, 'al-chem' means 'black soil'. The word means, refers to, the black soils of Egypt. Religions are like huge forests that take root in the ground of tradition, but religions—let us not forget—they come and go, and the soil remains. The roots of all traditional thinkers are in common soil.

'Alchemy' is an Arabic word, isn't it? It refers to 'black soil'?

Etymologically it refers to the black soils of the Nile valley. The rich soils of Egypt, along the Nile. And the color black. The grossest matter. That is the bed of the spirit. It is an important etymology. It refers to the soil of *tradition*. The soil in which our religions, the world-religions, have taken root. Our religions are over-growths. Underneath is the bed, the soil. *The black soil is the earthly correlative of the darkness beyond the stars.* This is speaking metaphorically, but in practice the farmer's vocation is a *way*, and actually one that suits the temperament of far more people than does the vocation of priest or monk. In modern suburbia we witness countless people hurry home from their jobs, their abstract, meaningless jobs, and rush out to catch a few hours of daylight in their gardens: this is symptomatic of a longing for the earth among the earthy, a picture of a peasantry displaced to the suburbs and the factories. Work-a-day lives are preoccupied with abstractions; for a great many people there is a need, however unconscious, to put their hands into the earth to touch reality again. Wilderness is a symbol of divine agency in creation, a mighty force that places Man in context, but Eden was a *garden*, a symbol of Man and God in harmony.

So religions, you think, are like 'crops' in the soil? And 'tradition' is the soil? Doesn't this make religion less important?

In an important sense religions are *secondary*. Tradition is the soil. Religions grow. Religions die. In the beginning no one needed religion. Remember—*in the beginning no one needed religion.* They are relatively new things in the life of man. This is something that religionists forget. Religions are superstructure constructed over earlier foundations. It is common these days to emphasize the extent to which the agricultural classes in medieval Europe remained largely untouched by Christianity or else blended a smattering of faintly understood Christian motifs with older, pagan practices. True. But this is only to be expected. Tradition comes before and outlasts this or that religion; the agricultural classes, people of the soil, are the true repository of the perennial wisdom that is beyond the particularisms of the various religious orders. The Sufi brotherhoods insist that their traditions are timeless, that there were Sufis *before* Islam. True. They mean by this that the brotherhoods are linked intimately with and embody the wisdom of the traditional crafts, which certainly existed before Islam. The Sufis attach themselves to the Islamic revelation, but true wisdom—in a very real sense—resides with the blacksmith and the woolcarder. And these crafts came before Islam. And of all traditional crafts, farming, working the land, is the most fundamental. The blacksmith serves the farmer. The farmer is first. The husbandman's participation in the processes of creation is prototypal in a traditional society. In a sense, the initiatory lines of the religious mysteries, with their selection of candidates, initiations, trials and chains of transmission, as well as the bloodlines of aristocracies, are *imitations* of the farmer's craft. Schisms, crises and heresies all come and go in the life of a religion, but a traditional civilization cannot stand for long the disruption of agricultural life where not only the basics of life are produced but a wisdom of the earth moves steadily from mouth to ear—even from breath to breath—across the centuries nurturing transient superstructures. This 'low' wisdom, if we can call it that, stems from the adequacy of traditional crafts as languages, systems, of symbols of the divine. The religious mysteries are extensions of the mysteries of the land, the silent mysteries of the land. They are unspoken—even *unthought*—in the farmer. They live in the blood.

If no one needed religions in the beginning, as you say, why do they need them now?

Because we have become removed, abstracted, from nature,

from *the revelation of nature*. The revelation of the *kosmos*. The cosmic revelation. In the course of things we lose sight of it. We forget it. We forget how to read it. We become blind to the text of nature. We need to appreciate how utterly abstracted and alienated is the human condition. And the character of that alienation is not just creature from Creator but creature from Creator *and* Creation. Then you need to appreciate what religions are for. Their purpose is to reorient man, to *re-vivify the text of nature*, to render it intelligible again. Religions are adjustments. Radical adjustments brought about, made necessary, by human alienation. Nature is the primal revelation. Nature is God's book. In the beginning, primarily, the *kosmos* is translucid, a complete revelation. Because *the kosmos is a mirror of Heaven*. But what happens, as I understand it, is that man becomes abstracted from primal revelation and as he falls away from it the Principle 'responds' with a series of reiterations appropriate to the deteriorating conditions. First there are the primal traditions, then at last the scriptural ones. Primal traditions are prior to the scriptural traditions and the scriptural traditions are already a falling away from a more integral state and the primal traditions are nearer to it.

By 'primal' traditions you mean so-called 'primitive'? Like the Australian aborigines, for example?

Except—you must understand—these traditions, as we know them, are just remnants. They lost their integrity long, long ago. Even before contact with Europeans. We only know the remnants now. But in their integrity primal traditions are—were— superior to scriptural ones. In some ways, though, this is just to say that oral traditions are prior to written ones and in that sense superior to them. Eventually, these reiterations 'solidify' into written forms. It's a solidification. Written revelation comes last in the cycle and only comes about when men have lost all sense of the 'translucid *kosmos*' and no longer understand its 'existential language'. The conspicuous thing about religions is that they *point us back to nature*. They compensate for what we have forgotten. They teach us how to read the primal revelation, the book of the *kosmos*, again. Observe the Quranic injunction: 'Read His signs on the horizon and in yourselves!' There are *three* things that must be reconciled: Creator, creation, creature. Islam, at least, is opposed to an abstracted Otherworldliness. The Qur'an is as nature-directed as much as God-directed. The whole message of the Qur'an is that God's signs are staring you in the face. In the stars. In the clouds. The air. The dusk. The dawn. In the smell of the desert after rain. In your *body*. All translucid *revelations*. But how to read them? Well, for a start, modern people labour under extraordinary abstraction from nature. We can't read the signs—the *ayat*, in Arabic—of nature because we have literally lost the grammar. Cosmology is a grammar.

And in Christianity?

The subject of the sermon on the mount isn't the Law of Moses, its nature. Consider the lily, Christ says. He points back to nature. Consider the lily in the field. Learn from the sparrow. He says, *'Learn to read the primordial text again.'* The more grounded, the more lived, your cosmological sense, the more lucid will be your scripture, whatever it is, Bible, Qur'an, Vedas. Islam never loses sight of how important it is that a human being be immersed in a network of meaningful symbols. The Qur'an, if you will, is like a book of slogans that you can stick onto the things around you like stickers. Sacred art is nothing more. A set of labels to stick onto nature to remind us of realities. It is as if we were living in a world where we had forgotten the names for everything. So God sent us a print-out of name-tags. *Revelations supplement nature in this way.* They seek to restore our primal relationship to Creation and Creator. That's what religions are for. If we had never lost our sense of the 'translucid *kosmos*' we would never have needed religions.

But aren't religions whole revelations, complete in themselves? How can they be secondary in the way you suggest?

True. We need to avoid suggesting that the later scriptural traditions are in some way derivative from the primal traditions. No tradition depends upon another for its *principle*; rather every 'new' tradition consists of *renewed contact* with the Principle. However, there is a sense in which the scriptural traditions 'amend' the primal ones or rather 'compensate' for them or 'correct' our relationship to them. For example, as fallen man loses sight of what is 'translucid' in nature and tends more and more to worship nature itself—by which the integral primal traditions become 'paganism' properly defined—scriptural traditions assert the Creator over the creation. Just to set things right again. It becomes necessary to 'remind' man that nature is not God, but only 'signs' of God. The Qur'an tells us that the whole of Creation is a language of signs (*ayat*) and implores us to read it. Nature is not to be worshipped but *to be read as revelation*. A constant Quranic theme. But the Qur'an itself does not instruct in the grammar of the text. The purpose of the Qur'an is to *contextualize* the text by reasserting the singular transcendence of its Author, so that we are in the position to read it correctly to start with. But for the grammar and wherewithal of reading we must turn to the traditional cosmological sciences. Every aspect of the scriptural traditions can be seen as secondary to the primal ones. Among the Red Indians, for example—in such a primal tradition—*every man is a prophet*. It is as though every soul is prophetic. Every soul has a prophetic nobility. Scriptural traditions, on the other hand, which come along later, represent a *particularizing* of the prophetic man.

And the Jewish tradition?

Similarly, only when man ceases to be 'naturally' righteous does it become necessary to express the Principle as 'Law'. Law is a falling away from a pristine condition *beyond* law. There is a general principle at work. Scriptural traditions are a type of 'involution' of the primal revelation. The Book is Nature made *microcosmic*. The Book is a microcosm. It is *extracted*. The Book is *parallel* to nature. Thus throughout the traditions we find the equation, Nature = text. The Torah—like the Qur'an—is *kosmos as text*. Involuted as text. The *kosmos* has been collected up, extracted, and rendered as a text, a book. That is the essence of the Jewish tradition. The Torah is a microcosm. Man has become removed from nature, so nature is *involuted* as a text, collected up, made in a *tincture*, essentialized as text. The Torah is the universe as text.

You mean the Torah contains the whole creation in principle?

Yes. The Qur'an is the same. The Qur'an is especially clear on this. The interweaving between the Book and Nature is striking in the Qur'an. The two are parallel. There is a sense that what is revealed in nature is now—*perforce*—being made into Book. (The historical 'cause' being the arrival of literacy among the desert Arabs). And the Qur'an, like no other Scripture, is replete with accounts of the things of nature as revelations from and of God. This is a characteristic of the Qur'an as the 'final scripture'—it points back to nature as the primal revelation. The Qur'an embraces all previous scriptures but, importantly, iden-tifies nature—translucid nature—as their base. Scriptural revela-tions are always *additional* to nature, the primal revelation. What they do is *correct* our relation to that revelation, not introduce a different revelation from nothing. All scriptures refer to the scriptures that came before them. And all scriptures refer to nature as the primal revelation. But, all the same, scriptural tra-ditions are in no way derived *from* nature, but are revelations with the same authority. The change of revelation from nature to revelation from scripture was not via nature but *by a renewed contact with the Principle*. Thus in Quranic terms God has a Divine Tablet and Divine Ink and a Divine Pen, and the whole technol-ogy of literacy is sanctified.

But earlier you said that divine wisdom comes from the soil, not from the pen.

The struggle is to sanctify—incorporate—new technologies. To make them sacred. To make man-made technologies sacred as they arise. This is why, ultimately, we become divorced from the sacred. Because we cannot keep up with technological change. When man discovered, controlled, fire, then fire became a sacred symbol. It was written into the *code*, if you like. And when man invented the wheel, the wheel became a sacred symbol. Written into the code. And when man developed the plough, the plough became a symbol. And when writing was invented, literacy, then new revelations took that form and that new technology was sanctified, naturalized, made native to the *kosmos*. Made *sacred*. But this is still a falling away into greater and greater abstractions and the encoding becomes less and less convincing. Until finally—historically we can notice it towards the end of the Middle Ages—we just cannot keep up. Technology gets away. And soon man is surrounded by a profane world of profane things that obscure rather than illuminate the translucid *kosmos*. That is how decline occurs. Cyclic decline. Eventually we cannot make things into symbols. One of the main tasks of the Islamic revelation was to make the technology of writing into symbols, sacred symbols. To sanctify writing. But for all of that, you see, the Prophet was *illiterate*. Unlettered. And the Qur'an does not 'draw upon' nor is *derived from* nature's revelation, but is a Revelation in itself—in a different mode—and, as such, is 'equal to' nature's revelation, which is to say *has the authority to amend to advise upon it.* The Qur'an, by appeal to the same Authority, amends and comments upon the primal revelation, nature itself. It is a 'force equal to nature'—but nevertheless secondary. Scripture becomes necessary because of our increasingly disturbed relation to the primal revelation (nature). For example, we tend to worship the things of nature and forget their Creator. Thus was the Qur'an necessary to correct man's relation to the primal revelation and remind him that nature is not God but 'signs' of God. Or, more broadly, thus were the monotheist scriptures, beginning with the Torah, necessary to ensure the One wasn't forgotten in the many. Similar corrections to our relation to a primal position are the stuff of all scripture. Even when they appear anti-nature, scriptures are correcting our relation to nature as primal revelation, and it will be because we are too immersed in nature *to see nature as symbol*. Scripture sometimes replaces nature, and can do so, but only temporarily and in an *emergency* as it were. Thus as man lost contact with his Principle and ceased to be 'naturally' righteous, Law was needed. But Law was a falling away. And religion was needed and is a falling away.

PART FOUR

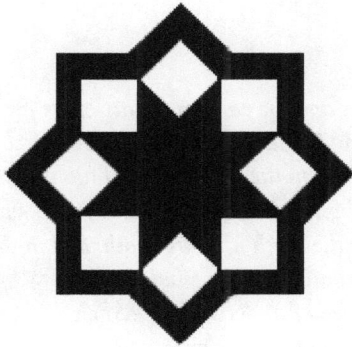

Surely in the heavens and the earth there are signs for the faithful: in your own creation, and in the beasts that are scattered far and near, signs for true believers; in the alternation of night and day, in the sustenance God sends down from heaven with which He resurrects the earth after its death, and in the marshalling of the winds, signs for men of understanding. Such are God's revelations.
Qur'an 45:2–7

Cyclic Compensation in Islam, Christianity and Judaism

AS the last revelation and the last of the world religions Islam is tailored to the conditions of cyclic decline, which conditions include the decline in the spiritual, mental and other qualities of men. Islam addresses these conditions by providing a revelation of greater clarity and simplicity than those that came before it and, within the context of that fresh revelation, a mitigation of earlier severities. Most obviously, Islam simplifies and clarifies the Abrahamic/Judaic complex of religious tradition while at the same time reducing the severity of such strictures as Judaic food laws. This is a Quranic theme: The End approaches and so We are making things both clear and easy for you! Moreover, both the Qur'an and the Traditions of the Prophet have an awareness of the hastening of decline and cater to the fact that, as time goes on, even the mitigated clarity of Islam cannot prevent a decent into confusion and irreligion. One of the most important characteristics of the Islamic revelation is that, while it is a return to the primitive simplicity of primal revelation, it presents this in an inherently flexible form suitable to the conditions with which it must contend, namely the conditions at the end of the cycle where men are far from the integrity of primal man. In the context of the whole cycle it is a religion for weaklings. According to the Traditions, God had originally required fifty prayers from men each day, and the whole Qur'an had to be recited at each prayer. The angels praise God continually: God requires less of man, given that he is a creature of clay. But the Prophet—acting as Intercessor—was ashamed to represent to God that his people could not possibly endure such a burden. God mitigated the requirement to forty prayers. Still the Prophet said that this was something his people could not bear. So God mitigated the requirement to thirty prayers, and so on, until at last it was reduced to five prayers per day after which, we are told, Muhammad was too ashamed to plead for further mitigation. At the same time, the requirement that the whole revelation be recited at each prayer was reduced to the recitation of the Fatihah which, by God's Mercy, contains the whole Qur'an in principle and so fulfils the requirement in a manner within the capabilities of men. But even then, we find other Traditions explaining that, as the End hastens, five prayers is too much and that fewer and fewer Muslims will maintain the prayer but that God, ever-Merciful, will make one prayer worth a hundred in those times and that the rewards of maintaining the prayer increase as conditions conductive to maintaining the prayer decline. These Traditions are characterized by the ideas of proportion and compensation. Typically, for each step man takes to God, so God takes first ten, then a hundred, then a thousand steps to man, or runs to him ever-faster, as it becomes harder and harder for men to take that one step to Him. Man was created weak, and his weaknesses grow and dominate him the more remote he becomes from his primal state, but his Creator is cognizant of this and compensates accordingly, and Islam is proof of this for it is primal revelation reconfigured by God in His Mercy for the benefit of the spiritually puny creatures into which Man has hardened. Islam is preeminently a religion of cyclic compensation.

Christianity is also a religion of the End Times and therefore also a religion of cyclic compensation, but in a very different

manner. For a start, it does not entail mitigation of former strictures in the same way as Islam; rather its compensatory nature
takes an entirely different form and it specifically rejects the
Islamic response to cyclic decline. The Two Commandments of
Christ are not a mitigation of the Decalogue but rather an *exposure of their interior*. The baptism of John replaced the burden-
some purifications of Second Temple Judaism with a single
washing, but this was to reveal their inner significance, not to
make things easier for men. Similarly, Paul did not seek to abol-
ish the practice of circumcision to make conversion easier for
Gentiles, but rather to externalize the 'circumcision of the
heart' of the Torah. Christ came not to overthrow the Law and
the Prophets but to *fulfill* them. In this, it is not mitigation that
Christ brings, as James and the so-called Ebionite Christians
supposed, but, as Paul maintains against them, *freedom* from the
yoke of the Law altogether, which is to say, in Christianity's Jewish context, the exteriorization of *an inherently esoteric freedom*.
This is the important thing to realize about Christianity: it represents an inherently esoteric viewpoint of which the exoteric
form is, properly speaking, Mosaic Judaism. The compensation
to cyclic decline, in this case, takes the form of exteriorizing an
interior perspective (namely Edenic freedom) so as to make
accessible to a broader range of souls Mysteries which had for-
merly been accessible to only the few. This was in response to
the rapid hardening of cyclic conditions under the Roman
Empire and the dramatic decline of the sanctity of not only the
ancient Temples and Oracles and Mystery Schools but, more
importantly, of *exoteric* sanctity. Christianity—especially in its
central resurrection myth—was an exteriorization of what had
previously been an esoteric, initiatic order in the ancient world.
But the veil was rent asunder. All was exposed. In the Christian
perspective, God's response to cyclic decline was to sacrifice His
only begotten Son in an incarnate, pivotal dramatization of the
death and resurrection Mysteries in order to arrest *exoteric*
decay. The Mysteries themselves were in decline and their
integrity could no longer hold, primarily because the exoteric
supports that are essential for the maintenance of such inner

Mysteries had effectively collapsed. The inner, we must remember, cannot long survive without the protective shell of the outer, and in the Christian case we see the phenomenon of an inner perspective reconfigured—perforce—as an outer perspective in order to offer the prospect of *salvation* to as many souls as possible as the end of the cycle nears. This is not a case of mitigating an exoteric Law but of invoking a perspective that is beyond Law and so intrinsically internal. The proportionate compensation of the Divine to the conditions of Late Antiquity was to throw open the Mysteries to all and to install a sacred pivot into the hardening historical (and anti-mythological) worldview fostered by both the Romans and the Jews. The death and resurrection of Christ, fixed into a single historical event and a single *avatar*, serves the cyclic purpose of making the experience of the Mysteries accessible to all not by *gnostic* initiation but by *bhaktic* identification. In doing so, of course, Christianity imperilled the integrity of the esotericism it brought out into the open, as well as displacing Judaism, and necessarily manifested a tendency that is against non-exposed esotericisms, a tendency that in fact eventually hardened into a formal position. Instead, the proper inner dimension of Christianity is cloistered and monastic and is *removed from the world* and only non-exposed in *that* sense.

The issue here is the distinction between spiritual liberation and mere salvation. In a traditional order, the exoteric religion provides a means to salvation while the concern of the esoteric dimension of that same religion is complete liberation even from that blessed state called 'salvation'. As a consequence of cyclic decline, as the quality and strength of souls diminishes, salvation becomes almost as difficult as liberation had been in former times, while liberation itself becomes proportionately more difficult and rare. The simple fact is that men are not the beings they used to be and even the most rudimentary spiritual discipline becomes too much for them. And on top of man's diminished inner powers, his whole environment conspires against any form of spiritual realization. Furthermore, the nature of the cycle is that the Iron Age serves as a preparation for the Golden Age of the next cycle so that salvation of souls rather than their liberation from the cycle altogether is, in a sense, the proper work of the Iron Age—it is the Golden Age that offers the conditions that make liberation most accessible; such conditions decline as the cycle proceeds. In such circumstances, the Mercy of the Divine is directed in the Last Days to the salvation of the many rather than the liberation of the few and this is precisely a *leitmotif* of Christianity. It is significant that traditional Christianity takes a monastic form and imposes the monastic ideal upon the Roman Empire. This represents a narrowing of the path to liberation and is, as it were, a counter-balance to the extension and broadening of the path to salvation. As the exterior of Christianity is already an exposure of previously esoteric symbols and doctrines, its interior modes become more remote and difficult to attain. The Christian settlement is that Christ makes salvation easier—He died for our sins and passed through the Initiatic death for us—but at the same time liberation—where there is none but Christ and no 'me'—now requires the extra ardor and mortifications of fully monastic vocation. Liberation is not by any means impossible within the Christian orbit, but in cyclic terms liberation is not its primary or original focus, and indeed it may be said that, more than other religions, it neglects and is sometimes hostile to this dimension in favour of a universal salvationism in which, nevertheless, liberation is accessible to those that can attain it by virtue of the intrinsically esoteric and inner nature of the Christian symbolism. In this respect, in Christianity—pre-eminently the religion of the Holy Spirit—it is far more important that the 'spirit bloweth where it listeth' than in other religions.

A contrast between Islam and Christianity should be noted at this point: in (Sunni) Islam the exoteric shell is anti-priestly and denies priestly modes of succession, while the Sufi *tariqah* are rigorously apostolic, forming chains (*silsillah*) back to the Companions and to the Holy Prophet himself. In Christianity the reverse tends to be true. It is the external aspect of the religion that finds authority in apostolic chains of transmission, while there is no continuous, formal initiatory esoteric structure. Liberation, in

Christianity, takes the form of *mysticism* in the proper sense, usually in a monastic context. The heresy of the Templars (and the Masons after them) consisted of attempting to install into the Christian order a Sufi-like (indeed, Sufi inspired) tariqah-style Christian esotericism, in defiance of the contrast we have just noted. While the strongly Sufi-Judaic spirituality of the Carmelites could be absorbed into Latin Christendom at the close of the Crusades (though not without causing significant disruption), that of the Templars could not. The difference was that the Carmelites could be situated within the monastic structures of Christianity, which is to say that they could be absorbed into the monastic core and made to conform to monastic norms. The Templars, on the other hand, proposed importing a new type of esotericism—fraternal but not monastic—that was in direct imitation of the Solomonic Sufi esotericism of the Muslim Near East. At a later date, under the cover of the Protestant Revolt, Rosicrucianism attempted to do the same. These were cases of failed cross-faith appropriations and adaptations—attempts at an Islamicized Christianity—not signs of a 'secret' Christian tradition. In the present writer's view, much Traditionalist opinion on these issues—opinion usually formed in the context of the modern French encounter with Islam and its 'occultism'—is colored by the assumption that a Christian esotericism should take a similar form to Muslim esotericism, namely by means of initiatic brotherhoods. This is to misunderstand the very structure of the Christian religion. If there are signs of such a type of esotericism in the gospels, specifically in Johannine texts, it should be observed that there are no signs of monasticism: the esotericism of the early Christians over-lapped with the ancient Mysteries somewhat and even among the Church Fathers (although only among those later condemned as heretics, such as Clement of Alexandria) we find allusions to certain 'secret' teachings, but Christian orthodoxy eventually made a different settlement and the remainder of these 'mysteries' were extinguished with the pagan ones or woven into monasticism, especially in its strongly Johannine Eastern forms and specifically in such works as those ascribed to Dionysius the Areopagite. To distinguish here between different types of Christianity, the forms of which we speak are most acute in Latin Petrine Christianity where externalization fixes upon the person of the Pope, an exoteric priest, while the monastic core is more integrally developed in the East. Needless to say, Protestantism is a rupture and a sudden decline because it further exteriorizes the esoteric by emphasizing faith alone over faith and works—Luther's was an esoteric point of view—and at the same time abolishes monasticism leaving the religion with no coherent core whatsoever. Nevertheless, in the terms we have considered Christianity here, Protestantism can hardly be called an illegitimate expression of the Christian spirit but rather, as it purports to be, faithful to the first impulse of the faith, for it is very much an extension of Christianity's cyclic *leitmotif.* In Protestantism we find what are often extremely esoteric viewpoints laid bare, in no proper context, and treated in the most externalistic ways, and both salvation and liberation are *really* reduced to a cases of the 'spirit blowing where it listeth', which is to say to more or less 'chance' conjunctions of factors only made possible because of the inviolable status and powers of the symbols themselves. If it seems we have just portrayed Christianity as a one-dimensional exotericism, it is because this is not an inaccurate description, provided we remember that its symbolic content is anything but exoteric in itself.

It needs to be appreciated that Christianity, like Buddhism, is a special case and an *extreme* correction to cyclic conditions. Islam, like Hinduism, is more 'normal' in this respect and so a *stabilization*—a return to 'normal' or primal patterns—made necessary, in part, by the disequilibrium created by the advent of Christianity which is, in a sense, a tradition that is *inside out,* so to speak, with the interior exposed and the exterior (i.e., Judaism) locked in ghettoes. More generally, it needs to be appreciated that, as cyclic conditions decline, certain compensations need to be made and that one of these is that salvation becomes a more urgent matter than liberation. The Sufis rightly regard Paradise as a prison but the fact remains that one needs to get to Paradise before one can escape from it and of this modern man is less and less assured. In the case of Islam, cyclic compensations take the form of proportionate adjustments—a 'flattening'—until a new equilibrium is restored, but even then its ultimate concern is for the salvation of the many—it 'flattens' or 'broadens' Jewish Law to do this—and thus, outside its core, it is a *mass* religion in which believers swarm to Allah like insects. Modern Islam, infected by Wahhabism and subject to other pressures, steadily betrays or forgets the Sufi tradition and increasingly hardens into an empty externalism, and an externalism of merely external symbols—Muslim externalism is *all the more external* because of this. That is the fate of Islam. The fate of Christianity has been for its monastic core to shrivel and its exoteric unity to crumble, leaving the spectacle of the most profound of esoteric symbolisms falling into the hands of the very worst of men. Moreover, it needs to be said that, as an exposed esotericism, Christianity is inherently disruptive and cannot tolerate the esoteric dimensions of other religions it encounters—it is an agent of cyclic exposure. All of this amounts to saying that the religions of man are becoming progressively more *shallow* and *horizontal* and are losing sight of their inner dimension, and Christianity's inherently esoteric order of symbolism certainly does not save it from this but rather makes it an agent of such shallowing, as it challenges other traditions to empty their inner contents into the stark daylight of history too. Thus its viciously exclusivist stance. Islam can accommodate Christianity because it can enwrap the exposed esotericism of Christianity in a new exterior—the protective shell of Islamic Law—and by the device of tolerance for 'People of the Book' re-establish or re-permit Christianity's esoteric relationship to Judaism. But Christianity can offer no reciprocal accommodation to Islam, (and no compromise to the Jews either) not least because its whole perspective disrupts and defies the concentric inner/outer structure of traditional Islam. In declaring a universal freedom from the Law Christianity also necessarily declares the distinction between the Law and the Way, the external Law and its internal realization, irrelevant, making the entire inner/outer model void.

As far as Judaism is concerned, it is not a question of cyclic compensations but of being overtaken by cyclic events. The vigorous proselytizing Judaism of the Second Temple that challenged Romanism with its 'ethical monotheism' was splintered, yielding Christianity, firstly, and secondly a retarded and ossified rabbinism almost entirely occupied with externalist praxis. In this respect, the Jews were *victims* of cyclic compensation, and their 'adjustment' to cyclic conditions involved the utter destruction of the Temple, its cultus and its priestly lines and a corresponding retreat into a separatist, bookish legalism that surrenders its remaining esoteric possibilities to the device of Elijah's return. If an esoteric philosophy, namely the Qabbalah, blossomed in medieval Judaism, and if continuations of Second Temple themes such as the Merkabah mysteries found new inspiration, they mainly did so, it must be said, under the stimulation of Moorish culture and in an incomplete, syncretist form, and, in any case, the Qabbalah notoriously led its adepts to embrace Christianity, acknowledging that the inner dimension of Judaism *is* the Christian faith after all. Similarly, other forms of Jewish mysticism dwell upon the identity of man and God and the anthropomorphic body of God, nascent Christian themes. Nevertheless, because the external symbolism of Judaism—the Law—is unmitigated and unadjusted it remains, in a sense, more symbolically translucid than the Islamic *Shariah*—the Hebrew alphabet is a more primal set of symbols than the Arabic for all of Islam's primitive spirit—and hence is inherently 'deeper' or more metaphysically perspicuous, and so, even in its externalism, Judaism remains a potentially fertile spiritual order with an inner dimension that may, in theory, be accessed via quite outward symbols and practices, although with correspondingly greater dangers of diabolical miscalculation, it should be said. The 'chains of transmission' in Judaism are, like Elijah—nay, *with* Elijah—occluded but exist as spiritual realities accessible by Eliatic transport made possible by the Divine Presence in the sacred text. But there are no compensatory devices in Judaism beyond the end of the obligations of the Temple regime. Reform and 'Liberal' Judaism are human contrivances of unsanctioned mitigations in the Islamic style, we might say. In fact, the severities of the Torah still hold even though cyclic conditions are against the narrowly exclusivist salvationism of the Jews these severities maintain. Nor is there a New Covenant to open sanctity to the mass of End Time souls, but only a continuation of an embattled and, in our times, belligerent exotericism which may even be described as counter-cyclic. Without mitigation or other compensations Judaism remains encumbered with a rigor of which its adherents and the weak souls who populate the cycle's close are increasingly unworthy. The 'esotericism' of Judaism, we might also say, consists in the status of the Jews as 'Chosen People' among all men, which gives their exoteric order an esoteric quality insofar as they can say that their exotericism is subject to a special Divine favour—*among exotericisms*. The whole concern of rabbinic Judaism is the identification of the Jews as the 'Chosen People', which is to say the Golden Race of the forthcoming cycle (of whom circumcision and resurrection of the dead are emblems), and so liberationism—by which one seeks to escape the cycle and avoid this destiny—may even be construed as a *sin* in rabbinic Judaism. It is not going too far to say, referring to levels of symbolism at least, that Judaism is an *esoteric exotericism*, while Christianity is an *exoteric esotericism*, and in this way they form a necessary complement to each other.

There is no prospect of this complementarianism ever reconstituting a new unity, however, except in an apocalyptic context, but it is entirely to be expected that these perspectives together form a bloc against Islam such as we see in the contemporary context. The effect of this on Islam—sustained militarism in its Middle Eastern heartland—has been—and will continue to be— a hardening of Islamic externalism and the further marginalization of Sufism which—like monasticism in the modern Christian environment—seems like escapism in the face of on-going

civil and political crises. It is no accident, given all we have discussed so far, that the locus of Jewish ambitions—in league with
those of certain Christians—is the Temple Mount which, in
Islam, is the Quranic *Farthest Mosque* and the site of the Prophet's
Night Journey, which is to say the precise symbolic axis of
Islam's 'vertical' or esoteric dimension. There can be no mistake
that the confrontation of these three religions in the Holy Land
in our times serves the end of eradicating exactly this aspect of
Islam. The Third Temple movement in Israel envisages demol-
ishing the al-Asqa mosque and rebuilding the Temple, reconstituting the ancient priesthood and recommencing the ancient
sacrificial order. Nothing could better symbolize the cyclic processes that are afoot, for this would amount to replacing the living symbolic center of Islamic esotericism with an exoteric idol,
a monument to empty literalism and the power of finance. We
cannot avoid stating that, in cyclic terms, the prolongation of
Judaism beyond the advent of both Christianity and Islam
increasingly becomes an *irritant* that acts as a catalyst and focus
for cyclic decline. The Christian-sponsored return of the Jews
to the Holy Land is an event of cyclic importance, but any
notion that this represents a 'revival' of Judaism is mistaken for
there can be no substantive 'revival' of Judaism until Elijah
cometh at the very end of the cycle, Zionism being exactly
heretical on this point. The state of Israel represents nothing
other than a hardening of Judaism into a concrete Biblical literalism of purely political scope. Jewish spirituality becomes inex-
tricably bound up in this earthly, nuclear-armed project to
reduce the religion's symbols to 'facts on the ground'. Certainly,
Judaism is not the same as Zionism but who can deny that the
state of Israel has consumed Judaism and made conditions such
that Jews can hardly focus their attention elsewhere, and that its
concrete existence makes a purely inner and symbolic, spiritual-
ized reading of the Hebrew Scriptures all the more difficult, for
contemporary 'facts on the ground' intrude into almost every
narrative on every page.

The reason for the prolongation of Judaism, which, properly
speaking, should have been fully superseded by the Christian

religion in the 'normal' course of events, is precisely because of the extreme and peculiar nature of the Christian adjustment. Had Christianity merely mitigated Judaism in the Ebionite manner, there would have been no basis for the Jewish prolongation, but because Christianity adopted a more extreme mode of cyclic adjustment—namely exteriorizing an intrinsically inner perspective—Judaism necessarily had to continue because Christianity did not *replace* it in a directly one-to-one manner. While the Christians become the New Israel, the old Israel is nevertheless 'un-negated', to coin a term, and persists if for no other reason than to remind Christianity of what Christian 'freedom' is freedom from. Judaism is the counter-force in the cycle. And it is furthermore because this matter was left unresolved, and so a radical disequilibrium was created, that the third revelation, Islam, was necessary at all and when it appeared it did so in the semitic environment. The intrusion of Islam— unexpected and unaccountable as far as the Christian perspective was concerned—creates its own internal stabilization, in which the earlier religions, Christianity and Judaism, may participate—Moorish Spain being a conspicuous instance of this—but at the same time the presence of the primal recapitulation that is Islam must also introduce a new level of disruption simply by adding a third term to the equation. In fact, this disruption eventually proved fatal to the Christian order as it traded its spiritual integrity for the means to compete with the 'Infidels' on a material plane and Christian heresy—the whole of Protestantism included—gathered around such pseudo-Islamic themes as the 'priesthood of believers'. And, as in the case of the Templars and Rosicrucians, 'occultism' in the Christian environment very often took (and takes) similar pseudo-Islamic forms, even to the point that most so-called 'magical words' in Western occultism are garblings from Arabic, the Templar's 'Baphomet' being the infamous and paradigmatic case.

These, in conclusion, are the cyclic configurations that dominate our times and as the end of the cycle hastens they become all the more stark and apparent. It needs to be added that, while the religions we have been considering are, by definition, devices to postpone cyclic decline and to provide shelter from the storm, they are not by any means immune from cyclic decay themselves and are, from another point of view, not only symptoms of decay but agents of it as well. We must remember, for example, that these 'Religions of the Book' already represent a hardening of revelation into a literate form and have together waged a contemptuous war against pre-literate traditions—let us mention the Amerindians—who were indiscriminately labeled 'pagan'. For all that Islam is an expression of Divine Mercy to men in these Last Days, and for all its official policy of tolerance towards 'People of the Book' it is, in other respects, a menacing force hostile to integral sacred perspectives that do not obviously conform to the narrow definitions of semitic monotheism; most notable is its hostility towards Hinduism, Islam being, as it were, that religion's cyclic nemesis. So the intrusion of Islam, with all its compensations, into the cycle provides, as it were, a life-raft for souls in the coming flood, but it also further complicates cyclic conditions and, by the very nature of things, does damage as well as good. This is certainly true of Christianity as well. Part of its mission, for example, is to sacralize time and the historical consciousness, and while this is a blessing and a mercy, it then neglects or is hostile to the mythological mind and so draws a wider and wider orbit of people (including the Muslims who followed them chronologically) into a historical mentality. More significantly, its posture of exteriorizing an inherently internal point of view disrupts the 'normal' inner/outer distinction, and so while it is an agent for universal salvation it increasingly takes specifically anti-liberationist or anti-esoteric forms insisting that all that is inner must be laid bare and contextualized by merely sentimental devotions. To adapt a traditional symbolism to this: the exteriorizing of Christianity is like a cyclic 'descent' by which cyclic conditions become more *shallow*, while Islam is like cyclic 'amplitude' by which cyclic conditions become *flatter*, and Judaism is like the 'weight' or 'gravity' that ensures these two movements are destructive rather than creative, declining rather than regenerating. These things, we should add finally, are never a question

of one religion being superior to another but of simply understanding their distinctive natures and seeing their place in the
economy of the whole cycle and its current phase. The position
of the present writer is, that, nevertheless, when all things are
considered, only Islam has a properly constituted Way to liberation (as opposed to salvation) and a proper (primal) relationship
between inner and outer, and this clarity of levels is its privilege
as the last revelation: both Christianity and Judaism are, in
themselves, a rupturing of this arrangement such that Christianity is a core that has been exposed and Judaism is a shell that
awaits a new core, and in neither the Christian or Jewish environments has there ever been a formal, legitimate 'inner path' to
liberation comparable to the Islamic *tariqah* because a formal
inner/outer model does not prevail in those environments by
definition. This is not to say that liberation is impossible in those
environments but it exists in very different modes and, in any
case, like all End Time religions—Islam included—their focus is
necessarily on *salvation* and preparation for the new cycle. It is
disingenuous of Traditionalists and others to complain that they
cannot find initiatory modes of spiritual training in Christianity
and Judaism such as one finds in Hinduism, Islam and the Mystery institutions of the ancient world—that is not at all how
cyclic compensation is configured.

Autochthony
and the Symbolism
of Islamic Prayer

IN some respects *salat* or the canonical Islamic prayer ritual is the most important of the five essential observances called the Five Pillars of Islam, for it is the only one to which the believer is required to adhere every day. The Testimony of Faith is the most essential of these pillars in absolute terms—without it one is simply not a Muslim—but Testimony is only required once in a lifetime, as is the Pilgrimage to Mecca and in the case of the Pilgrimage it is only if circumstances permit. The Fast of Ramadan and the Poor Tax are annual requirements, and for these there are exceptions. The prayer, however, is daily and there are no exceptions except the allowances made for menstruating women. In the practical life of the Muslim, therefore, the *salat* is the greatest burden of the Law. The Qur'an, strictly speaking, only petitions believers to practice 'regular prayer' (most often, 'keep regular prayer and pay the poor tax') but in the codifications of Islamic Law this settled into five canonical times per day: at first light, at noon, midnoon, after sunset and at night. Not only is this cycle a personal discipline and a means by which the believer punctuates the day with worship, it also has a symbolic significance that illuminates the deepest roots of the Islamic faith. So too do the prescribed bodily postures and movements of the prayer ritual which are themselves cyclic in nature—*salat* is measured in terms of cycles or *rakas*—and are part of the same symbolism. This symbolism concerns the primordial myth of cyclic regeneration. As the final revelation and at the same time a reiteration of the primordial revelation, Islam has the cosmic cycle—and especially its approaching end—as a central theme. The theme takes symbolic form in the cycles of the daily prayer times and, more explicitly, in the cycles of the prayer gestures and movements.

As far as the cycle of prayer times is concerned, it is based on the daily movements—risings, settings and culminations—of the Sun and provides Islam with its particular development of solar symbolism. In Christianity and other religions solar symbolism is expressed in terms of the Sun's annual cycle—its passage between the equinoxes and the solstices—but in Islam the annual cycle is determined by lunar movements not solar and instead it is the daily cycle of the Sun from east to west and from the heavens to the underworld to which Islamic spirituality is attuned. In this, the daily movements of the Sun repeat, within twenty-four hours, the same cycles it completes in a year so that dawn and dusk are parallel to the equinoxes and noon and midnight to the solstices. Even though Islam assiduously avoids even the impression that the Sun itself is an object of worship and so, for example, forbids prayer during the actual risings and settings of the Sun (noting, however, that, conversely, it provides special prayers to mark an eclipse of the Sun in a directly helio-focused way), the prayer times are nevertheless arranged around the two axes of the Sun's diurnal movement—the horizontal and, in the greater cycle, equinoctial, axis of East-West and the vertical, solstitial axis of Up-Down. In the Christian perspective, the two axes of the equinoxes and the solstices introduces a cruciform symbolism that develops the identification of Christ with the annual death and resurrection of the Sun. Islam, of course, makes no such development but the symbolism of the solar cross is nevertheless marked by the prayer times; opposite to the dawn prayer is the dusk prayer and opposite to noon prayer is the night prayer. The fifth prayer time in this arrangement, asr, represents a projection of the center of this cross (the quintessence) and thus is marked for special attention: it is the 'middle prayer' that the Qur'an specifically yet cryptically adjures Muslims not to neglect, the designation 'middle' referring to its centrality, not to it being in the 'middle of the afternoon' as externalists will commonly explain. Despite there being five prayer times—and this, with the Five Pillars, gives Islam the five-pointed star as one of its emblems—the symbolism of the prayer times is essentially axial and fourfold (and thus is related more to the symbolism of the *Kaaba* than to the Five Pillars) and marks the cross of the four extremes of the Sun's daily course which, in turn, are the same four extremes marked by equinoxes and solstices in the annual cycle. By extension, both daily and annual solar cycles reiterate the still greater cycles of cosmic time including the cycle of the Four Ages which cycle is near its completion as the very existence of Islam signals. We need only note that the primordial perspective of Islam—in which the duality of night and day is more essential than the quaternity of the seasons—tends to see the two solar axes as separate rather than developing them together into cruciform symbolism as does Christianity.

Throughout Islamic symbolism the two axes marked by the Sun are presented as expressions of the fundamental distinction between the notions of 'deputy' (*khalifah*) and 'slave' (*abd*). The Islamic understanding of the human condition is founded upon these two interdependent, indeed axial, ideas. On the one hand, according to the Islamic view, man is a deputy of God on earth. He is appointed as God's *representative,* and as such he has responsibility for the earth and its resources. The all-important correlative of this, however, is that man is also the *slave* of God and as such is utterly dependent upon Him and is bound to obey His laws and to exercise the right of deputation according to God's revealed wishes. In his status as *khalifah* Islam conceives of man as a free, noble and self-responsible agent, but in his slavery to God this freedom is held in check, man is kept from a self-deceiving Prometheanism, and he must confess to himself and to his God his mere creaturehood. This amounts to the distinctive Islamic formulation of the Biblical doctrine of stewardship, for man *controls* but does not *own* his earthly habitation. In axial terms, the dignity of God's *khalifah* denotes a vertical symbolism while the humility of His *abd* is horizontal by contrast. Prayer in Islam is always conceived as *dhikr* or 'remembrance' and in the cycles of the canonical prayer the worshipper is reminded of exactly the deputy-slave distinction (*khalifah-abd*) and exactly this understanding of the human place in the Creation. This is not achieved by functional or liturgical differences between the various daily prayers, however. The format and content is the same in every case. There is no special ritual for the dawn prayer or the night prayer or the others. The only thing that distinguishes the two axes is the number of prayer cycles (*rakas*) to be made at each juncture: in this the prayers of the dawn-dusk axis are comparatively brief (two and three *rakas* respectively)— while the prayers of the noon-night axis are longer (both four *rakas* each), which difference reflects the relative velocity of the Sun at the equinoxes compared to the solstices and also the varying lengths of the Four Ages. But there is no difference between one *raka* and another. The *raka* is the basic unit—one complete cycle of prayer. And so the structure of the *raka* is the primary mnemonic by which the believer is reminded that he is both the deputy and the slave of God, and it is in the structure of the *raka*, not in distinctions between various prayer times, that the axial symbolism sketched above is most explicit.

There are several worshipful postures of the human body admitted in Islamic ritual practice, but two of them are of fundamental significance: standing (*qiyam*) and prostration (*sajda*). The prayer begins in the standing posture. The Muslim faces the direction of the *kaaba* in Mecca—which is to say the *qibla*—in an upright but relaxed standing position, with the feet slightly apart and the hands either folded near the navel, over the breast or hanging free naturally, according to minor divergencies between different schools of thought. It is in this position that the Muslim is to be conscious of being God's *khalifah*. This is clear from the liturgy recited at this point:

> You alone do we worship, and to You alone do we turn for help. Guide us to the straight path...

This is the *fatihah* or Exordium to the Qur'an which is conspicuous among the Quranic revelations in being a prayer directed to God, and which is moreover conspicuous for being in the *plural*. The Muslim does not pray to God in this context in terms of I, me or mine. Rather, the *fatihah* is a collective prayer and standing in this position, symbolically facing his Lord, the Muslim represents not only himself but all mankind and even all Creation as *khalif*. In his capacity as God's deputy, man is also intercessor, and in the standing position of prayer the Muslim stands as intercessor to God on behalf of his fellow creatures. The vertical, standing posture is unique to man and betokens his unique position over and above all creatures: his uprightness confers upon him a distinguished place in creation. It is the physical correlative to the status of man as pole between the various ontological realms, between the Earth and the Heavens, between God and God's created order. The standing position with which the prayer ritual begins, therefore, fulfills the vertical symbolism of the two axes we are considering.

The prayer then proceeds—via an intermediary bowing posture which itself is highly suggestive of a cross configuration and which is in fact indicative of the *Asr* or 'middle' prayer time—to *sajda* in which the Muslim drops to hands and knees and presses his forehead to the earth. This is the quintessential gesture of submission and obedience in Islam which at once acknowledges God's supremacy over man and man's subservience under God. It is the gesture of surrender and as the mystics of Islam describe it of *fana* or annihilation of the ego before the Divine. It thus expresses man's slavery, for in prostrating himself the Muslim acknowledges God's absolute Lordship. The annihilation of the ego is further symbolized in *sajda* by the fact that the face of the worshipper is hidden from view in this position; the surrendering of all selfhood is expressed, throughout the Islamic tradition, in the veiling of the face. One of the strictures of Islamic art, for instance, is a prohibition on portraying the Holy Prophet's face. This is because the Prophet is submitted to God paradigmatically and is, as it were, always in *sajda*. *Sajda* is a direct, physical expression of homage, allegiance, fealty, compliance and service. It is by this posture that the worshipper is reminded of his lowly status as creature and of Islam's demand of man that he relinquish all pretensions to self-sufficiency—pretensions that may too readily accompany the status of deputy—and place himself unreservedly and faithfully, like a dutiful servant, in the hands of his Master. That the head is pressed to the earth brings the Muslim to the horizontal plane, and the symbolism of the second of the axes we are considering comes into play. The horizontal axis is receptive, submissive and passive as opposed to the active and dynamic vertical axis. In prayer, the Muslim moves from the vertical position, signifying man as *khalif*, to the horizontal, signifying *abd*. A cycle of prayer or *raka* is constituted of sequences of movements from the standing posture, to *sajda*, and is completed when the worshipper returns to the standing position again.

In Islam's own account of its spiritual heritage the *salat* is a restoration of the ancient Abrahamic mode of worship. In the Traditions of the Prophet, the Angel of the Quranic revelation, *Jibreel*, taught the movements of the prayer to Muhammad, instructing him that this is the way that his ancestor Abraham prayed. The prayer, therefore, is part of Islam's claim of returning to the pure Abrahamic faith. The symbolism of the prayer, however, is more primordial than Abraham for it must be remembered that Abraham is the *post-diluvian* founder of the *semitic* monotheisms and was himself restoring a former order from the depravity of his father's paganism; the prayer ritual instituted by Abraham was not unique to him but only a new codification—appropriate to the new post-diluvian conditions— of yet older, indeed, primordial modes. In truth, the *salat*, in its essential aspects, is the prayer of Adam, just as Islam was his religion. This is obvious since Islam acknowledges that there were prophets before Abraham and does not hesitate to say that these prophets all prayed to the same God in the same manner. In mythological and visionary formulations of Islamic prophetology Muhammad, as the 'seal' of the prophets, is *imam* (prayer-leader) to all previous prophets back to Adam and they pray the *salat* together. That Islam traces its heritage to Abraham is the specific means by which it situates itself within the semitic family, but beyond the semitic family and before Abraham was Noah and before Noah was Enoch and before Enoch Adam, and all of these were prophets and all prayed the prayer of the prophets. Abraham was responsible for the particular adaptation of the *salat* after the Flood in the context of his covenant with God, the emblem of which is, in Islam as in Judaism, circumcision. But Abraham was not the first prophet and the primordialism of Islam is not confined to the semitic family but extends back to the ante-diluvian prophets and specifically to the first prophet, Adam himself. In any case, while the prayer ritual is ascribed to Abraham its basic symbolism is certainly more primordial and is not, in the first instance, Abrahamic itself. The only explicitly Abrahamic motif in *salat* comes at the end when, in the canonical liturgy, the worshipper—raising his index finger to the *qibla*—calls a blessing upon Muhammad and his family and upon Abraham and his. In other respects, the whole symbolism of the prayer is Adamic. As we have seen, the principle movements enact the doctrine of stewardship and the essential relation of man as *khalifah* and *abd* to God and the Creation—all of this concerns Adam as primordial man and God's steward, not (at least in the first instance) Abraham as father of the semites. If the particular liturgical formulation of the prayer employed by Islam is Abrahamic, it is nevertheless the Abrahamic formulation of a more primordial, Adamic, symbology, and we have identified its core symbolism, namely the axial contrast of deputy and slave.

While the legal postures—standing, bowing, prostration—of the prayer ritual obviously enact the gestures of a slave doing obeisance to his Master, at a more profound level they also enact the story of Adam. In the standing posture, facing the universal center, the *qibla*, the Muslim is Adam in the sense of *everyman* and in the liturgy he speaks as Adam for all creatures and all creation. The corresponding symbolism of the *sajda* then becomes clear. In the prostrate posture, with face occluded and forehead to the earth—the paradigmatic gesture of the whole spirituality of Islam—the Muslim is Adam returned to the passive clay from which he was created. Here we must remember that while Islam, even more than Judaic and Christian cosmogonies, insists on *creatio ex nihilo*, which the Qur'an presents in the starkest terms—

> Creator of the heavens and earth! When He decrees a thing, He need only say Be! and it is! (Kor. 2:117)

—it also absorbs the primordial Adamic mythology in which God fashions, in a demiurgic manner, the primordial man. 'We are the clay, You are the potter, we are all the work of Your hand!' said Isaiah (64:8) and similarly Jeremiah, 'As clay is in the potter's hand, so are you in mine, O House of Israel' (18:5) both of which reiterate such passages from Genesis as: 'Of dust you are and to dust you shall return.' On these matters the Quranic account is no different:

> He is the Mighty One, the Merciful, who excelled in the creation of all things. He first created man from clay.... (Kor. 36:7–8)

When the Muslim presses his forehead to the earth in submission, he is making the Adamic confession that he was born of this earth and that he is but as clay in the hands of his Creator. The Qur'an has a persistent fascination with Adam's lowly origins. Although Islam counts Adam as the first of God's Prophets, and although the angels were made to acknowledge him, and although the Islamic perspective absolves Adam from the responsibility of an original sin and presents him as God's appointed deputy, he is nevertheless born of the common clay of the Earth, as the word Adam itself signifies, and the Qur'an dwells upon his ignoble origins. It is because Adam is born of mere clay that Satan (Iblis) rebels, refusing to acknowledge a creature of such low birth. In the symbolism we are considering, however, the stuff from which Adam was made—earth, clay, dust—is essentially malleable and submissive to the craftsman's will and so is an emblem of submission to God. In the standing position, the Muslim is Adam as God's deputy over creation, but the humility of *sajda* reminds us of Adam's origins and the primal, submissive state of the Earth from which he was crafted. Moreover, the movement of the worshipper from the vertical to the horizontal and back to the vertical, constituting a single *raka,* enacts the birth of Adam from the soil. In *sajda* he is malleable clay. In *qiyam* he is the upright form Allah has molded from the clay.

The notion that man is earthborn or autochthonous, which is to say made of the very same soil he tills and over which he labours, upon which he lives and in which he is finally buried, is a core tenet of the universal doctrine of cosmic cycles. The story of Adam is both the Biblical and Islamic autochthony. In the craftsman analogy that underpins it clay is a *prima materia* that offers no resistance to the shaping hands of the craftsman God. It bends entirely to His will and is shaped entirely according to His design. And just as there is an axial paradox in the human state, given that this lowly slave is also dignified by the title vice-regent and stands over all other creatures, so Adam is given as his domain the very Earth from which he was formed. And it is the fate of all the children of Adam—for the Qur'an also dwells upon his status as 'ancestor of all' quite aside from Abraham as 'ancestor of Arabs'—to return in death to the Earth and, at the end of time, to re-emerge resurrected. In the myth of cosmic cycles, more explicit in traditions less adapted to the impending climax of a cycle, the primordial man was born from the Earth *like a plant*, and his body is *like a seed* that shall grow anew, be resurrected, at the commencement of a new cycle. Adam is an earth-man, not explicitly a plant-man, but the plant motif is suggested by the 'Tree in the Midst' with which he is associated; this is an important assimilation made in Jewish mysticism where the body of *Adam Kadmon* is superimposed upon the Tree of Life and more dramatically in Christianity where Christ is hung on the same Tree. In these traditions primordial man and primordial plant are usually separate, though man is made of the soil in which the tree grows, and there is always an implicit interchangeability between one symbol and the other. In traditions focused upon the climax of the current cycle we find a linear perspective that envisages a single trajectory from creation to Last Judgement. Islam is such a tradition, but more than either Judaism or Christianity, it is aware of the greater cycle, which is again an instance of its inherent primordiality. There are both verses of the Qur'an and many *hadith* that speak of cycles of creation, even if the immediate concern of Islam is this current cycle. Certainly, in Islamic sources, the children of Adam are said to rise from their graves like plants in a *vegetative* cycle of births, as is explicit in Traditions such as the following:

> Abu Hurairah relates that the Holy Prophet, upon whom be peace, said: Everything in the human body disappears except the little bone at the end of the spine from which its second creation is compounded. Then Allah will send down rain from heaven and people will be grown like vegetables.

This is recorded by both Bukhari and Muslim, the two most authoritative editors of the *hadith* literature. The resurrection is here presented in terms of the vegetation cycle. Man is made of Earth, born of the Earth, but is also *plant-like* in this respect and so at the resurrection rises from the dead *like a plant*. In other perspectives the primordial plant-man rises from the Earth at each new creation in a cycle of creations and resurrections. In particular, it is usual that the Iron Age lays the seeds for a new Golden Age. In the Islamic perspective, the Paradise or Garden that awaits believers is a renewed Eden, a return to the Edenic state, a cyclic return to the primordial garden. In Eden, of course, Adam was truly a native, an aboriginal, an autochthon, born of the soil. His punishment for transgression was to lose his home and to be sent into exile and to have to be naturalized elsewhere. An Adamic *homesickness* for Paradise is the characteristic mood of all Islamic spirituality. Salvation is a *return*. The Islamic perspective, *finally*, is cyclic. And Allah is He to whom the returners return.

Much of the symbolism of the Muslim prayer illustrates these themes, especially by way of the axial, Adamic contrast of *khalifah* and *abd*. The cycle of standing and prostration in a single *raka* illustrates a vegetative cycle; it enacts the birth of Adam and ritualizes his dual status as deputy and slave, but the same gestures and movements also rehearse the death and rebirth of the resurrection of the sons of Adam. *Sajda*, then, symbolizes death. By the same symbolism the prayer mat symbolizes the grave and for this reason the traditional and symbolically correct design for the prayer mat is a stylized Eden of four rivers with the Tree of Life, the destination to which the believer aspires and for which his soul yearns. More significantly, it will be noticed that the position of *sajda*, with forehead to the ground, and arms and legs rounded into the body, is distinctly *embryonic* or *foetal* in form and suggests the embryonic seed sleeping in the grave awaiting the Last Judgement when 'Allah will send down rain from heaven...' It is far easier to see this when the worshipper is dressed in traditional Islamic costume which is loose fitting and designed to facilitate the movements of the *salat*, not hinder them as does modern Western clothing. The traditional robes dissolve the specific outlines of the human form—itself indicative of the notion of returning to the malleable passivity of the clay of Adam—and in *sajda* they make the human form seem foetal, consistent with the order of symbolism we are exploring. The return to the vertical axis in the prayer then enacts birth (and resurrection), the whole movement signifying the cyclic creation/re-creation of the primordial man. In the bowing posture that is between standing and prostration Muslims will typically elongate their backs so as to create a right-angle with their bodies. This underlines, as we said, the integral cross symbolism, but it should also be noted that, in fact, the consequence of stretching the back and forming such an angle with the back in this way is to mark *the bone at the base of the spine*—named in the *hadith* as the seed of resurrection—as the corner of the right-angle. That this is what is being emphasized is apparent if one witnesses a pious Muslim at prayer straining to perform every gesture in the exact, prescribed manner. The secret to performing the bowing posture correctly is that one shapes the back to emphasize the base of the spine. Other details of the ritual make similar allusions to the themes of cyclic regeneration.

Autochthony is also a theme of Abrahamic religion, but adapted to the conditions after the Flood. Like Adam, Abraham is bound to leave his actual ancestral home—the land to which he has a connection by birth—and to find a home elsewhere. There, in the Promised Land, he must form a legitimate connection with a new home, a new soil. The immigrant cannot enjoy true and natural autochthony but in order to be natural*ized* they must acquire an autochthony of sorts, and this is what happens in the story of Abraham. Thereafter, the claim of being son of Abraham becomes synonymous with a claim to the (acquired) autochthony of the patriarch. The equation expressing the autochthony becomes—

> I will make your descendants like the dust on the ground: when men succeed in counting the specks of dust on the ground, then they will be able to count your descendants. Come, travel through the length and breadth of the land, for I mean to give it to you.

—by which the descendants of Abraham are identified with the dust and soil of the land, each descendant like a speck of dust, a speck of the land itself. Circumcision, which is the outward sign by which the promise and covenant are made, is here, as in other traditions, directly emblematic of autochthony by way of primordial serpent symbolism, for the penis is serpentine (an Edenic symbolism) and the serpent is reborn from its hibernation in the Earth each year and the foreskin represents the skin it sheds each spring: to remove the foreskin is a mark of the new-born, the serpent in spring, and so also an emblem of the Golden Race who are the new-born of the great cycle. This motif, when adopted as a racial extension from an illustrious ancestor, is appropriately marked on the organs of generation and in Judaism at least becomes the mark of a 'Chosen People'. We see then the full sense of why the *salat* is attributed to Abraham in Islam, namely that the story of Abraham is about the acquisition of autochthony and so reiterates exactly these otherwise Adamic themes.

A further important symbolism emerges in the story of Abraham, too, for the revelation of the equation of his descendents with the specks of dust on the ground occurs in a dream, before which, we are told, Abraham 'fell into a deep sleep'. This is of great importance in Islamic prophetology, for the deep sleep state is the ground of prophecy, even in the case of Muhammad who—according to the *hadith*—would suddenly drop into deep slumber (and even snore) when receiving the Quranic revelation. Deep sleep is analogous to the prophetic state—to the Unlettered purity of Muhammad, in Islam—in its pure passivity. By this analogy we return again to the axial symbolism of *khalifah* and *abd*, for the waking, conscious mind of man accompanies his uprightness of posture, while he returns to the horizontal plane to sleep, and the waking mind is indicative of his dignity as *khalif* while the opposite to the waking mind, the mind in deep sleep, is perfectly passive, *abd*. The standing posture of the prayer therefore symbolizes the waking consciousness, the bowing posture the dreaming mind, and *sajda* the mind in deep sleep. Further, the mind in deep sleep is, by this chain of associations, equivalent to the clay, the *prima materia*, from which the primordial man was made. In the Abrahamic autochthony it is *the sleep of Abraham* that is *actually* the soil from which his descendants spring. In *salat*, *sajda* is the point at which the worshipper seeks to utterly submit to his Lord, to annihilate the false ego, to be perfectly submissive to Allah's every impression, like clay, like a prophet, like the sleeping mind, like an obedient slave to God.

This finally brings us to a point that needs to be made concerning not just the postures of prayer but also the prayer times. The Qur'an adjures the faithful to maintain vigils of prayer throughout the night and in Islamic symbology in general the dark night and not sunlit day is spiritually fertile and significant. This is consistent with Islam's general lunar symbolism. In practice, even the dawn and dusk prayers are made in darkness because dawn is really first light (defined by a piece of black cotton distinguished from the night, which occurs well before sunrise, and which allows for only shadowy forms) and dusk is really after the full appearance of the stars and the end of solar shadows. In the voluntary prayers given in the Muslim canon there are long cycles of *rakas* through the night and throughout the holy month of Ramadan the pious pray through the night, amongst other ways the fast reverses the normal flow of diurnal and nocturnal life. This is because the so-called 'dark Sun', the

'Sun at midnight', is central to Islam's core. By this we mean *the union of waking consciousness with the sleeping mind*, or in other terms, *conscious submission*. This is the very essence of Islam's lunar symbolism. The Moon is nothing less than the Sun of the night and a symbol of the waking mind united with the complete passivity of the sleeping mind. This again is what is enacted in the *salat*. Time and time again, cycle after cycle, in five sets per day, the believer, standing in *qiyam*, falls to the floor in *sajda*, which movement enacts *carrying the waking state to the sleeping state*. The Muslim's life is punctuated with this *dhikr*. Time and again the believer, in *sajda*, face occluded and forehead to the passive earth, foetal-like,—more often than not praying in darkness because of the organization of the prayer times—is, as it were, an image of one *conscious in deep sleep*. With the formulaic cry of 'God is Most Great'—the *takbir*, *'Allahu akbar'*—which expresses the transcendent, incomparability of the Divine and as a ritual formula dissolves the created, relative world before the Absolute, the worshipper, plunges, as it were, into the depths of the sleeping mind, namely that part of ourselves that is perpetually in submission to Allah and offers no resistance whatsoever to His Will. The method of Muslim prayer is just this: to consciously identify oneself with this deepest stratum of oneself that is, by nature, in perennial submission, to find in ourselves again the very 'clay' of which we are made. This is the deeper and a specifically Abrahamic dimension of the Muslim rite.

To recap: we have here sketched a series a parallelisms all of which are active in the symbolism of Islamic prayer. Both the prayer times and the *rakas* of the canonical prayer ritual rehearse both astronomical and cosmic cycles. The essential organization of the prayer times is fourfold, but in other respects Islam resists cruciform developments, emphasizing axial contrast instead. The essential contrast is between *khalifah* and *abd*, deputy and slave, which contrast is enacted in the two primary postures of prayer, standing and prostration. While the *salat* is ostensibly a restoration of the ancient Abrahamic prayer, these postures enact primordial and Edenic, Adamic themes. Specifically, the position of *sajda*—the fullest expression of submission—by placing the forehead to the earth, symbolically equates with the *prima materia* from which Adam was made. Further, the prayer ritual becomes, in certain of its movements, a symbolic enactment of the birth of Adam, and in this we find the particular Biblical and Islamic expression of the creation of the primordial, earthborn man. In other autochthonies the primordial man is explicitly a plant-man in a vegetative cycle of Ages, but in this tradition man is clay to God the Divine Potter, although the dead shall still rise from their graves like plants and the idea of the plant-man is implicit. Autochthony is also the concern of the Abrahamic dimensions of the prayer, but then by extension to the 'clay' of the sleeping mind. This in turn reveals the profound (lunar) theme of the 'Sun at midnight' in which the conscious mind seeks to embrace the primeval submission of the deep sleep state. In the specific linear perspective of the monotheisms, it is usual to speak of one Creation and one Judgement, but Islam is finally cyclic, for the resurrected return to the Edenic Paradise. Certain of the movements of the ritual also enact the embryonic dead in their graves and their resurrection into the after-life, the new cycle, growing up out of the Earth like plants. When the Muslim prays, all of these parallels of symbolism are activated, and by constant repetition, day after day, cycle after cycle, Islam hopes to actualize these symbols in the believer's soul.

The Mysteries of Wine & Spiritual Transformation in Sufi and Christian Perspectives

They are given to drink of a pure wine, sealed, whose seal is musk—for this let [all] those strive who strive for bliss—and mixed with water of Tasnim, a spring whence those brought near [to ALLAH] drink.

Qur'an 83:25–28

I tell you solemnly, I shall not drink any more wine until the day I drink the new wine in the kingdom of God.

Gospel of Mark 14:25

The authors of our being, remembering the command of their father when he bade them create the human race as good as they could ... placed in the liver the seat of revelation.

Plato, *Timaeus* 71d

He made me aware of an ancient secret...

The Qasida al-Khamriyya (Wine Ode)

Introduction

AT an external level one of the most obvious and palpable differences between Islamic and Western civilizations is that alcohol is prohibited in Islamic society while it is widely available and culturally acceptable in the West. Put plainly, alcohol is absent from Islamic social life while it is an altogether normal feature of Western social life—a stark contrast. This, of course, reflects the religious foundations of the respective civilizations: the Qur'an and the sunna of the Prophet Muhammad forbid wine-drinking while in Christianity—or at least in the orthodox branches of the faith—wine is nothing less than a sacrament.

This seems an irreconcilable difference, a matter about which there can be no rapprochement between the two civilizations and the two faiths, a point on which Islamic and occidental opinions seem diametrically opposed and on which there is no common ground.

It is important, therefore, to explore this divisive issue and to bring to it deeper, more universal perspectives, to rediscover and then to reiterate points of view and modes of understanding that have been lost and forgotten in our troubled and superficial times. As is the case with so many other issues that create Islam/West tensions the contrasts turn out to be not so stark, and at the deepest levels, in view of perennial and universal truths, they cease to be contrasts at all. There are dimensions to these religious traditions where we find that there is, in fact, a common platform and a convergence of doctrines, symbols and methods, and that what appears to be a contrast is really a complementarity Beyond their differences, the two religions are rooted in common mysteries and have overlapping paradigms.

In this paper we will examine something of these mysteries, not only as they are expressed in the current Sufi traditions of Islam and by Christian mystics but also in view of the ancient gnosis and wisdom sciences that came before the historical manifestations of either the Islamic or Christian spiritual orders. The symbolism of wine is primordial and extends into matters that are fundamental to all human spiritual life and to the entire spiritual constitution of man. For there to be truly meaningful dialogue between Christian and Islamic civilizations on this pivotal issue we need to recover a wider and more profound context of spiritual knowledge and begin to reacquire a lost understanding of ourselves as spiritual creatures, both before God and in the world.

A Symmetry of Paradoxes

There is a symmetry of paradoxes between Islam and Christianity on the issue of wine. In Islam, where wine is forbidden, the mystic, paradoxically, seeks a divine intoxication. Conversely, in Christianity, where wine is compulsory (because communion is an obligation), the mystic is often given to an ascetic abstinence. This fact should immediately suggest that the external contrasts are not absolute; the spiritual seekers of both faiths very often defy and push against the norms of the social order around them which is, in the process, exposed as only a shell, a protective covering.

There is a tendency among Muslim externalists—understandable and not illegitimate at is own level—to demonize alcohol in all its forms and to portray it as vile and evil, the urine of Iblis. Good Muslims will shun alcohol in every way. In popular Islam alcohol is often portrayed as a foul and sinister pollution. But the mystics of Islam, the Sufis—while adhering to the strict sunna of the Prophet—know that wine is paradisiacal in nature. It is *haram* not because it is infernal but, on the contrary, because it is divine and among those things in creation reserved for Allah. The Qur'an attests that there are rivers of wine in Paradise and declares that the denizens of Paradise are given pure wine to drink:

> They are given to drink of a pure wine, sealed, whose seal is musk—for this let [all] those strive who strive for bliss—and mixed with water of Tasnim, a spring whence those brought near [to ALLAH] drink.

The injunction for all those who seek bliss to strive to drink the pure wine of Paradise (forgoing the wine of this world) is the warrant for the Sufi's zeal for a sacred intoxication. There is a cult of divine inebriation in Islam, a drunkenness for God, although it is symbolic and even in that necessarily constrained by the social shell that surrounds it and to which it is a kernel. In the interior dimensions of Islam, extending from this warrant in the Qur'an, there is a doctrine of wine and a method of sacred

drunkenness even if abstinence and sobriety is the norm of the
Islamic social order. The Sufi drunkenness is not literal, of
course. The Sufi ideal is to be inwardly drunk and outwardly
sober—this is in fact supposed to be a tension integral to Islamic
society at large.

In Christianity, on the other hand, there is a tradition of mystical abstinence, although this is necessarily curtailed by the
obligatory nature of the Eucharist. Abstinence is an interior perspective in Christianity just as drunkenness is an interior perspective in Islam. We find illegitimate manifestations of
Christian abstinence among many Protestant sects—these are
often Christian imitations of Islam inasmuch as Protestantism,
with its emphasis on the Book and its bourgeois foundations, is a
Christian adaptation of the Islamic spiritual order—but it takes
a more legitimate form among orthodox ascetics, such as, for
instance, the Carmelite monks, the mystical tradition of such
luminaries as John of the Cross and Theresa of Avila. The primitive Carmelites shunned all wine until the Order was forced to
accept the chalice as part of their accommodation into Latin
orthodoxy. They retained their vegetarianism and their going
barefoot, but their asceticism of abstinence was limited by and
adjusted to the essentials of the orthodox sacraments in which
wine cannot be avoided.

Traditions of Christian abstinence, though muted, are built
upon the scriptural warrant quoted at the head of this article:

> 'I tell you solemnly, I shall not drink anymore wine until the day I
> drink the new wine in the kingdom of God.'

It is important to note that the Qur'an and the Gospels are at
one here: wine will flow abundantly in the Garden, in God's
coming Kingdom, and those who long to drink that wine will
refrain from wine in this life. The sacrament does not violate
this injunction, though, for, by virtue of the Real Presence, it is
the new wine of the kingdom. The wine of the Christian
Eucharist—the blood of Christ—is thus the very same wine that
the Qur'an describes as flowing freely in Paradise. There have
even been a few Christian mystics who refrained from all wine

but for the blood of Christ and who fasted taking no food or drink but the Lord's Supper. Here we have two different expressions of the same symbolism. The parallel is confirmed by the Qur'an's description of the wine of paradise being tempered with the waters of *Tasnim*: the wine of the Eucharist must be tempered with water too, not only because this was the ancient custom, but because this 'new wine of the Kingdom' is the blood of Christ and we are told in the Gospels that when Christ was wounded in his side two fluids, blood and water—the two elements, wine and water, of the liturgy—flowed from his body.

The Thirsty

The most obvious symbolism of wine concerns unity and multiplicity. The grape, which forms in bunches, is quite obviously emblematic of the Many and of the realm of multiplicity; wine then is the single essence, the principle of Unity, contained in the Many. But, just as important, the symbolism of wine concerns distinctions of immanence and transcendence, and it is this with which the mystical dimension is concerned. The problem, or rather the conundrum, that wine presents to mystics of either Islam or Christianity is this: wine is the preserve of the Hereafter, the coming kingdom, and yet it is an almost ubiquitous thing in the present life. In all but the coldest climates it is very hard not to make wine. The juice of any fruit will readily ferment. The windfall fruit below a tree will ferment of its own accord. The necessary yeasts are everywhere. They live on every breeze and breath of wind. Fermentation is a spontaneous capacity of nature. This, though, is exactly the point and is why wine is subject to this type of religious symbolism: God is not remote and His blessings are not postponed until another life. The natural fermentation processes that make wine are a symbol of how the transmutation of the soul is possible in this world. God is ever-present. His Creation is leavened with His spirit. His banquet is served. His wine can be tasted now; indeed, not only tasted but drunk to the dregs.

By temperament the mystic is not, as we might imagine, the most pious believer and the most devoted worshipper—rather the mystical temperament is, in an important sense, faithless and skeptical, impatient and doubting. For the mystic wants proof and he wants it now. The promise of paradise is not enough. The mystic will not wait until he is cold in his grave to be with God. He wants to be with God in this life. He wants the immediate experience of the sacred. He is thirstier than other men. He cannot wait until the Day of Judgment to enter Paradise and quaff its wine. He wants to drink his fill of the wine of Paradise in the flesh, in this present world, immediately. This is one of the reasons why the mystical zealot is not always welcome by his fellow believers and why exoteric forces in religions often, quite properly, impose constraints and limitations upon mystical modes of the faith. The mystic is often a threat to the equilibrium of a spiritual order, such orders existing, after all—at least in the case of religions like Christianity and Islam— for the salvation of all and not just the liberation of the few.

Strange to relate, it is the doubting disciple, Thomas, who gives us a canonical prototype for this temperament in Christian sources. He is the disciple who wants to touch the Risen Christ and who puts his fingers into Christ's wounds so as to be certain that Christ is risen in the flesh and is not a mere phantom. Christ had told the Magdalene in the garden 'Touch me not, for I have not yet ascended to the Father! (*Noli me tangere*)' but Thomas, of all those who witnessed the resurrection, needed the tangible experience, the corporeal certitude of touching. While this makes Thomas the model of the skeptic there is, nevertheless, a Thomasine gnosis where the mystic knows Christ as a tangible Reality. This is why the Thomas episode is only found in the most esoteric Gospel, the Gospel of John. It is Thomas who attests to the Real Presence which is the principle of Christian sacramentalism and which underpins the mysteries of transubstantiation.

In the iconography of the incredulity of Thomas the doubter inserts two fingers into Christ's wounded flesh. These two fingers represent the two natures of Christ, man and God, and in the act of inserting them into the transfigured flesh Thomas realizes that there is no division, no duality, but that the risen flesh is divine and human at once. There is no final distinction between a man's corpse and his ghost—the resurrected body is beyond any such duality. The mystical temperament is stirred by this fact. It means that we need not shed our fleshly form in order to know God. We need not pine for a distant heaven. In God there is no here and there, no past and future. If we attain to the Eternal Now all is transfigured, spirit and flesh both, as all opposites resolve into the Divine Unity.

In Islam there are some who simply must see God face to face. They burn with a desire to know the Real—really!—and nothing less will do. The Qur'an declares that God is nearer to us than our very life vein, and that wherever we turn our face, He is there. The mystic wants to know this, not just know of it. Islamic spirituality is often expressed in terms of nearness to God, and 'Presence' (*Hazrat*) is a title given to the saints. The mystic seeks the Real Presence, to be so near to God that one can touch Him and be touched by Him. God is no mere vapor, He is immanent Reality, Ever-present. There is an important *hadith* in which the Prophet, praying the eclipse prayer, is seen to reach out into space as if grasping at some invisible thing before him. After the prayer his companions ask him what he was doing. He explains that, for a moment, Paradise opened before his eyes and its fruits were so real, his vision so corporeal, that he reached out to grab at the heavenly fruit. 'And had I grasped one,' he said to his companions, 'you would have eaten from it as long as the world remains.' Paradise is more real than this life. Christianity and Islam share the doctrine of the bodily resurrection, resurrection in the flesh. Paradise is a transfigured reality, a transmuted physicality, not an otherworldly dream.

Wine as Revelation

Akin to this order of symbols is the use of wine as a metaphor for revelation, a symbol of the manifestation and incarnation of the divine, of how the divine can be present in this world. In Christianity, Christ himself is the revelation, the revealed Logos, God incarnate, God as flesh and blood. His blood is the wine of life. In the spiritual order of Islam, the holy Qur'an is the Logos and the words of the Qur'an are the draught of blessedness, sweet to the lips, intoxicating to the soul. It is in the words of the Qur'an that the Uncreated is made manifest. In Christianity, where the liturgy takes the form of a sacred meal, one imbibes the divine wine from the chalice in the sacrament. In Islam, which reverts to a more primordial mode of liturgy that is not sacramental as such and that requires no priesthood, one imbibes the divine wine in the *salat*, in the ritual recitation of the Qur'an, in which the living God is present. That is to say that the formal recitation of the Qur'an is, as it were, the 'sacrament' of Islam, not in the mode of a sacred meal but in a more direct mode of invocation. God is really present in the Christian sacrament—so attests the orthodox Christian tradition—and, in a different mode, God is really present in the Quranic recitation where the worshipper actualizes the Divine Reality in the spoken word and living breath. The Qur'an is to speech as wine is to water.

Wine as revelation is an especially pronounced theme in Sufi poetry. Bistami compares the sequence of prophets to the production of wine. The true seed, he says, was planted in Adam's time and it germinated in the times of Noah. In the time of Abraham it sent forth its branches and in the time of Moses the grapes set upon the vine. In the time of Jesus they ripened and then 'Muhammad's time' he says 'saw the pressing of the clear wine' the miraculous draught of transformation. This is at the same time a description of the spiritual journey. Each of the prophets represents a spiritual station until one is drinking in the company of the winemaster, Muhammad, the Seal of the Prophets.

This parallel between the two religions becomes clearer when we realize that the transformative principle in both cases is *anamnesis*, recollection. The life of this world is a forgetting and God's revelations are reminders from and of Him. Recollection is the actualizing principle in the Eucharist for Christ said 'Eat! drink! Do this in remembrance of me!' while in Islam the Qur'an, and its ritual recitation, is *dhikr'allah*, remembrance of God according to the Quranic covenant: Remember Me and I shall remember you. Spiritual realization in both faiths consists of the act of remembering. It is our dallying in vain pasts and idle futures that keeps us from the reality of the moment. All methods of mystical realization consist of remembering the Ever-present God, seeking and abiding in the Divine Presence. As this relates to wine, the old wine will cause us to forget—it reduces the body to a dead weight, the mind to a stupor and the tongue to a slur—while the new wine is a remembering, a sparkling lucidity, an uplifting joy that inspires the tongue to a heavenly eloquence. The Qur'an itself is *dhikr*—from the tongue of the supremely eloquent Prophet. Similarly, at Pentecost when the apostles were filled with the Holy Spirit and were enabled to speak in 'tongues'—a divine eloquence—people said 'They have had too much wine.'

Remembrance, *dhikr*, is, in particular, the great theme of the Sufis and of the Sufi brotherhoods, and their methods of remembering—principally the invocation of God's Name—are routinely compared to drinking the sacred wine. The Sufi lodges are often called 'taverns of the righteous' where spiritual seekers gather to 'carouse' and grow intoxicated upon God's revelations. Although there is no sacramental meal, certain devices in Islamic spiritual life nevertheless suggest the idea of revelation as nourishment. Amongst other things, the fast of Ramadan has this purpose. In Christianity the Lord's Supper sanctifies eating and drinking. In Islam, the fast does the same. The worshipper forgoes all food and drink during the daytime and prayer is their only nourishment. At night, eating and drinking—often of an expansive, festive character—is interspersed with the special devotions called *tarawih*, long sequences of voluntary prayers (a feast of prayer) to which the Sufis are especially devoted.

With each cycle of these prayers a portion of the Qur'an is recited, so it is in the extended *tarawih* prayers that the worshipper may recite large sections or all of the Qur'an during the sacred month. The Sufis are especially dedicated to the tarawih prayers because they lead directly to and are an arena for the *Laylat al-Qadr*, the mysterious 'night of power' or 'night of awe', the most mystical occasion in Islamic spirituality, the night that the Qur'an describes as better than a thousand months. The *Laylat al-Qadr*—the night upon which the Qur'an was revealed to the Prophet—is the climax of Muslim mystical life, the night (related in an axial manner to the Night of the Ascent) where the very principle of revelation is open to the saints, the night of the intoxicating ecstasy of Divine disclosure.

The Liver as Organ of Revelation

The mention of Thomas the Doubter earlier opens up another symbolism that is very obscure in our present times but that is essential to any full appreciation of all the issues before us. We must remember, as already stated, that as historical religions both Christianity and Islam are founded upon, and overlay, deeper, more ancient understandings from those times before the historical religions were even needed. To come to an adequate understanding of many religious matters we nowadays have to make an effort to reacquire ways of seeing and thinking and knowing that were once common and developed but which are now rare and neglected. In the present case we must place the symbolism of wine against a background of a primordial cosmological and metaphysical heritage. This is a more primal order of gnostic understanding. It is more concrete, more visceral, less abstract. It is preserved in the historical religions but often in ways that become less and less accessible to the increasingly abstracted historical man. It is entirely necessary that we introduce some salient aspects of it here.

In the depictions of the Johannine scene of doubting Thomas inserting his fingers into Christ's wounds, it is always the wound in Christ's side, whence flowed the eucharistic blood and water, rather than the wounds of his hands and feet, that Thomas is touching, and the wound in his side is always, in these depictions, on the right side of his torso. This is where the liver resides among the internal organs. As naturalism begins to prevail during the passage from medieval to renaissance Christian art, and the Christian mysteries are emptied into the public domain, representations of the scene become more lurid, culminating in Verocchio's statue and, later, Carravagio's almost shocking painting. But these representations make it clear in the most graphic way; Thomas' fingers—his two fingers, representing the two natures of Christ—intrude into the wound in Christ's right side where the liver is exposed. In a typical play of renaissance symbolism, Thomas is sometimes quite plainly both touching and pointing, the image being a document on esoteric anatomy for those with eyes to see it.

Turning, then, to this deeper and more ancient (we might call it shamanic) dimension, it is of cardinal importance—but entirely overlooked—to situate a consideration of the mysteries of wine alongside an investigation into the mysteries of the liver. The two things belong together—just as macrocosm and microcosm belong together—and cannot be separated. Wine is to drink. The fermentation of wine recalls the fact that processes of fermentation have been internalized in the digestive organs of the human microcosm. The fermentation of wine-making is essentially a predigestion, and the transformation of foodstuffs into nourishment (or inebriants into visions) is completed in the subtle alchemy of that most complete of all symbols, the human body.

In modern times we might still appreciate wine as a spiritual symbol but any spiritual significance of the liver—*hepato*—the organ that metabolizes alcohol and purifies blood, is entirely lost to us. Modern industrial medicine knows the liver as a murky swamp of catalysts and compounds, or sentimentally as the body's 'wonderful chemical factory', but we no longer have the slightest idea why ancient and even neolithic man regarded the liver as the seat of the soul. In traditional accounts, the liver is regarded as the very seat of anima and, more importantly for our present purposes, as the organ of revelation, an internal spiritual 'mirror'. Our ancestors gave to this organ the exalted role of 'portal of dreams', 'seat of divination', and 'eye of revelation'. Let us recall that hepatomancy—divination by an examination of the liver of a sacrificed animal—was almost as widespread as astrology in the ancient world. But modern man has not the slightest clue what all of this attention to the liver might be about nor the slightest inkling that the religious mysteries of wine are connected to it. It is necessary that we spell it out.

The important thing to know about the liver is that it is the microcosmic internalization of the cycles of night and day and it is therefore the organ of duality. It has two cycles, a day cycle and a night cycle, the night cycle being triggered when we sleep and stop consuming food or drink. The iconography of the doubting Thomas—who, incidentally, is Thomas the Twin—is all about duality. The two fingers of Thomas touching the place where blood and water flowed from Christ's (right) side could hardly be a plainer signal. Blood and water are, by extension, the cosmic elements fire and water, ignis and aqua, and represent the primal duality in elemental terms, while the two nature's of Christ concern the duality 'God and man' and Thomas' doubts are to do with the duality 'spirit/flesh'. The liver, in esoteric anatomy, is the microcosmic seat of the primal duality 'night/day' (and hence, waking/sleeping).

These connections are signaled not only in Christian art but they are quite clear in classical mythology. In Greek myth it is the demiurgic Hephaestus (the Greek version of Ptah, the Egyptian creator god) who blends wine and water in his chalice (*krater*) for the gods, and—part of the same mythology—binds Prometheus to the rock where Prometheus will have his liver pecked from his body every day, only to grow a new liver each night. Prometheus has his liver pecked out and restored in cycles; it is a defining condition of his bondage. It is the state of duality to which Prometheus is bound by the Wine-blender.

The other crucial thing to know about the liver is that, in traditional understandings, it is not just a filter in the body's drainage system but is a sense organ comparable to an inner eye. Traditional thinking, in fact, regards the liver and the eye as akin and connected. The eye is the other organ that has a day/night cycle. It opens and shuts with waking and sleeping in parallel, or rather counter to, the cycles of the liver. In the alternations of night and day, when the outer eye is open the inner eye is shut, and when the outer eye is shut the inner eye is open. The liver is the eye of dreams, the seat of inward vision, an inner mirror. As well, the finely knit, dense, shiny texture of the liver is remarkably similar to that of the eyeball. The connection is obvious in cases of liver disease which manifests as a yellowing of the whites of the eye. We might also recall, in our present context, that alcoholism produces blindness and that we commonly refer to someone being 'blind drunk'. The liver is an organ of vision, but it is the vision of our dreams, not our waking life.

Any method of spiritual transformation that is concerned with transcending the condition of duality and attaining a higher unity, any spirituality that conceives of the human predicament as bondage to a realm of duality, is likely to involve practices calculated to impact upon or manipulate the cycles of the liver and to preserve the clarity of this inner eye. Almost always there will be a preliminary process of purgation in order to cleanse the organ, polish its mirror, sharpen its sight and differentiate its cycles. These cycles can then be manipulated by the use of fasting (and fast-breaking) as means of controlling the organ's triggers. The organ can also be influenced by either drinking or refraining from alcohol.

In the symbolic order specific to Christianity, the duality of the liver (physically evident in the two lobes of the organ) is further divided into a fourfold symbolism where the organ is seen as consisting of four half-lobes. Christianity adopts a cruciform symbolism here. It is essentially the same as the basic symbolism used in astrology, and indeed the whole basis of hepatomancy is that the liver can be read in exactly the same manner as a horoscope according to two axes marking east and west, up and down. Wine, in this context, is a distillation of the year, the four seasons, the four directions. Thus is wine graded by vintage. In its flavors and nuances a wine retains the essence of a year. Christianity uses wine as a tincture of unity to overcome a four-fold multiplicity.

Practices

These ideas underpin both Christian and Islamic mystical praxis. Beneath the external differences that exist between Christian and Muslim teachings there is a common sub-stratum of pri-mordial wisdom science. Both the Christian Eucharist and the Muslim prohibition of alcohol must be seen in this context. Christianity employs wine to achieve certain inner transforma-tions while Islam achieves the same transformations without it. Christianity, we should recall at this point, specifically addressed the degeneration of the ancient mysteries in the Graeco-Roman world—the mysteries of Demeter and Dionysius—and so took a sacramental form since these were agricultural mysteries con-cerned with bread and wine. But the Christian redress of pagan decadence was, in the total economy of things, excessive, and this excess precipitated Islam which, again in the context of greater cycles, restored an even older, and nomadic, equilib-rium. In either case, though, the steps in transformation are the same: purgation, illumination, unity. And even though the Qur'an is 'clear' wine, which is to say it is not a physical liquid to be drunk like the Christian sacrament, it is still the liver that 'digests' it, so to speak. Islamic mystical practices have the same organic basis as Christian spiritual realization.

In the Christian case the wine in the Eucharist—in which is united spirit and matter, God and the world, Creator and cre-ation; in short, the two natures of Christ—'works' upon and through the liver. The Christian mystic prepares himself with fasting and purgation and forgoes all food and drink except for the flesh and blood of Christ. In this, the wine—the blood—acts upon the purged organ almost like a Hahnemannian potency or perhaps, to use a better known illustration, like silver nitrate over photographic film. It is the catalyst for a spiritual alchemy.

Islamic practices are even more calculated even though they do not involve actual physical wine. The whole secret of the Islamic mode of fasting is that it reverses the cycles of the liver, and that this is akin to reversing the cycles of night and day, sleeping and waking. As the fasting Muslim goes about his business during the day, his liver is triggered into its active mode because no food or drink has been ingested for hours. Normally, this occurs during sleep. During fasting it occurs while awake. The point of this is to bring the inner and outer eyes into alignment, into a common focus upon the Reality that is inner and outer, subject and object, both. Usually the inner eye opens when we sleep, but Islamic fasting reorders these cycles so that the inner eye is open while the Muslim is awake and conscious. Thus the Muslim has the experience of being fully awake while his body, so to speak, is sleeping. He is effectively conscious in sleep. The conscious mind and the dreaming mind are both awake at once.

This is a nascent spiritual state that prefigures the resurrected state beyond the opposites of flesh and spirit. In the resurrection all opposites are reconciled. It is the state symbolized by the eclipse, the resolution of the Sun/Moon duality—thus the import of the hadith mentioned earlier. It is the same state sometimes depicted as the 'midnight Sun', the union of day-consciousness with deep sleep or, more commonly, as the union of the opposites male/female in the mystical wedding. Islamic fasting is part of a spiritual science calculated to re-order our metabolism in a specific way, namely to reverse the normal cycles governed by the liver, to prepare us on the psycho-physical level for states of spiritual realization.

There is also a common gesture found in Islamic devotion that acts upon the liver, though no one understands it as such anymore. When the worshipper sits up from placing his forehead to the earth in the canonical prayers, and he sinks backwards upon his heels, placing his hands upon his thighs, it is sunna to fold the right foot in a peculiar manner that leaves the toes facing towards the qibla (Mecca). To achieve this the body must lean rightwards somewhat, stretching the left side of the torso but compacting and placing pressure upon the right side. It looks an odd and uncomfortable position to non-Muslims, and seems an unnaturally yogic contortion of the legs that makes sitting contrived and has no significance other than it is said to be how the Holy Prophet arranged his feet and legs at that juncture in the prayer.

This sitting position is very obviously designed to reposition and push upon the liver. Once it is pointed out to us it is as obvious as Thomas pointing to the liver of Christ in Christian iconography. We can stand at the rear of any mosque and watch it. To place the legs and feet in this way exerts a mild pressure upon the liver. It is almost a massaging of the liver done by titling the torso rightwards. If we still understood why the liver is the 'eye of dreams' and thought of it as something more than a filter in a drain, we would recognize this fact immediately; this position in the canonical prayers is undoubtedly designed to have a directly physical effect upon the liver, the organ of dualities. Prayer consists of recitation of the Qur'an, then this pressing on or massage of the liver during the sitting position that follows. Recite the Qur'an. Press on the liver. Recite the Qur'an. Press on the liver. It is almost as if the Prophet's practice was to hold himself in this position to sooth the liver after each imbibing of the 'clear wine' of God's Word.

As far as *salat* as a symbolic 'wine drinking' is concerned it is relevant to recall here that alcohol impairs our verticality. A drunken man staggers. He cannot stay upright. He loses that most human of characteristics, his uprightness. Eventually he falls down. After the Muslim, standing upright in prayer, imbibes the Qur'an he falls down on his face in prostration. After each recitation, made while standing, he falls down to the horizontal plane. The symbolism of wine concerns not only the Unity/multiplicity duality but also the spatial dichotomy, vertical/horizontal, that is the operative symbolism in Islamic prayer and its alternations between standing and prostration.

In view of all of this, consider, then, the cumulative effect of the fast of Ramadan. In the fasting itself the cycles of the liver are reversed and the worshipper assumes the state of conscious sleeping. He is held in that state for a whole month. And throughout there is the nightly feast of prayer, drinking deep draughts of the clear wine of God's sweet and intoxicating Word. In addition, at each of the extended *rakas* the worshipper, resting with his feet and legs as the Prophet did, places pressure on the liver. The devotions of the sacred month are designed in this way. As well, the month is traditionally divided into three sets of ten days, leading to Laylat al-Qadr on an unspecified night in the third set. These three sets correspond exactly to the processes: purgation, illumination, unity. The first set of ten days is purging. The sharpening of the body/soul duality is preparatory in all systems of gnostic attainment. Its organic correlative is the purging of the liver. The second set of ten days is illuminating— many people report vivid dreams beginning in the middle part of the fast. Then, in the third part of the fast there is the greatest potential for the unitive experience, the Night of Power, only attained by the highest souls and with the grace of God.

* * *

There is no space here to describe all of the more subtle processes of this alchemy and how they culminate in the mystic's experience of Unity, but it is necessary to add—it is an exceedingly important point—that the liver is, so to speak, the pulse of the heart expanded into the duality of night and day, and conversely, in the heart we find the diurnal cycle of the liver condensed into the pulse of the moment. This latter condensation corresponds to the work of the Sufi *dhikr*, the perpetual invocation of the Divine Name in the heart, and of the equivalent practices in Christianity, such as the *mnimi Theou* of hesychasm. The Divine Name encapsulates the whole revelation. The Divine

Name is the precious wine. Wine acts upon the liver but all of its higher associations are with the heart, *qalb*, and the oasis of the Moment that is between two heart beats.

Conclusion

The most important thing to be taken away from these explorations of alchemical transfiguration is a sense of the wealth of the esoteric traditions in Christianity and Islam and a corresponding sense of how woefully inadequate is any merely sociological consideration of contemporary religious debates.

We also begin to realize that meaningful dialogue must engage the representatives of the mystical schools of both religions. Externalism on its own will only create polarizations, entrench contrasts and obscure the common foundations. The social tensions between Christian and Muslim civilizations over their respective attitudes to alcohol, along with a whole range of problems (not the least of which is a worldwide epidemic of alcoholism, a disease of the modern soul) can only be overcome by rediscovering a transcending truth. The only solution to the contemporary impasse is to reacquire those things we ignored or threw away in our self-destructive rush for the superficial benefits and glamour of modernity, to start remembering those essential things we have so completely forgotten. The task of conducting dialogue between civilizations is much the same as the task of mystical realization itself, namely overcoming distinctions without destroying them.

We need the special wisdom of the Sufis and that of their Christian brothers and sisters, gathered together to share the new wine of God's revelations, if a genuine dialogue of substance is to take place. The realization that awaits contemporary man is that mystics have been right all along. The wine is served. It is at the banquet of Love, where all opposites meet and all dualities are resolved into the One, that the mystics of all faiths come together.

The Hoopoe, Zamzam
and the Kaaba

An Exploration of Symbolisms
of Revelation in Islam

And We have created above you seven paths...
and we send down from the sky water in measure,
and We give it lodging in the earth.
Qur'an 23:17–18

Introduction

IN Fariduddin Attar's famous Sufic allegory 'The Conference of
the Birds' a group of birds are led across seven valleys in search of
the fabled Simorgh and one by one they drop out of the journey
until only thirty birds (*si morgh* = thirty birds, in Persian) remain
and they realize that they themselves are what they have been
seeking. In this quest they are led by the hoopoe bird who, in
allegorical terms, represents the spiritual guide, the Sufi Sheikh,
who directs aspirants (*murids*) on a journey of self-realization.

The whole poem is based in Quranic symbolism. In the *Surah* called 'The Ants (27) we are introduced to the hoopoe bird in the context of the story of Solomon and the 'language of the birds':

> And Solomon was David's heir. And he said: O mankind! Lo! we have been taught the language of birds, and have been given abundance of all things. This surely is evident favour.... And he sought among the birds and said: How is it that I see not the hoopoe, or is he among the absent?... But he was not long in coming, and he said: 'I have found out a thing that you apprehend not, and I come to you from Sheba with sure tidings....' (27:16–22)

Attar's allegory is a development of the idea that it is the Sufis, the mystics of Islam, who know this mysterious 'language of the birds' of which Solomon speaks—and thus can the Sufis be allegorized as a flock of birds—and of the idea that, of all birds, it is the hoopoe who discovers what is hidden.

Modern readers might find these references quizzical to say the least. What is the language of the birds? Why does Attar attribute it allegorically to the Sufis? And what is so special about the hoopoe bird (also called the common lap-wing) that it deserves this status? Why is the hoopoe the spiritual guide? Why can a hoopoe bird find what others 'apprehend not'? The symbolism that is in play here, both in the Qur'an and in Attar's poem, is likely to escape the modern reader entirely. They are apt to detect no more than a quaint or romantic story and to remain unsuspecting as to any deeper levels of significance. It is therefore necessary to give a simple and clear exposition of the symbolism and to sketch some further, related considerations that might help elucidate important matters that are otherwise long forgotten in our times. A consideration of the meaning of the hoopoe bird, in fact, takes us into some of the most central symbolism of Islam.

Language of the Birds

René Guénon has revealed in one of his remarkable articles on traditional symbols, that the language of the birds—the underpinning metaphor here—is nothing other than the Names of

God and more specifically the recital of the Names of God in the Sufi practice of *dhikr*, sacred remembrance. The purpose of the Sufi fraternities is to gather for the remembrance of God by reciting His revealed Names, either in silent meditation or in chant. This is the essential method of Sufi spirituality and it is by *dhikr*—recital of the Names—that the Sufis seek nearness to God and self-realization. The Sufis have an esoteric understanding of the Qur'an, looking beyond the surface meanings of the text. In this case, when Solomon says:

> Lo! we have been taught the language of birds, and have been given abundance of all things. This surely is evident favour.

Sufic exegesis understands this to be a reference to the fact that God has revealed His Names to mankind, that these Names are a path to *spiritual* abundance, and that this is surely a favour and a blessing to man.

But on what basis is the identification of the language of the birds and the Names of God made? Herein is the importance of the hoopoe bird. The hoopoe bird is the key. It is only when we appreciate certain facts about the hoopoe bird that we can understand why the Divine Names might be called the 'language of the birds'.

And the main fact about the hoopoe bird is that it attains its fame among the ornithological symbols of Islam because of its distinctive call. As its English name suggests its call is a characteristic hoop sound—hooo—hooo—hooo. To Sufi ears, this call is an uncanny reproduction of one of the hallowed Names of God in Arabic, the Name Hu, which features in Sufi *dhikr* and recitals. Allah is the Name of the Divine Essence and the proper Name of God, but the Supreme Name—which is also His 'secret' or 'hidden' Name—is Hu (the pronoun He) by which Allah refers to Himself many times in the Qur'an. It is said that the secret Name is hidden in the Qur'an but also that it is quite obvious to those that know the secret—it is hidden by being everywhere. This hidden Name is the second person pronoun and many Sufi Orders employ it as a chant, often accompanied by certain modulations of the breath. The hoopoe bird is therefore sacred because it knows the Divine Name. Here is a bird—common and inconspicuous though it is—that knows the hidden Name of God—hooo—hooo—hooo—. Thus can the Names of God be described as the 'language of the birds' and thus can those men and women who recite the Names of God—the Sufis—be allegorized as birds. Thus too can the Sufi Sheikh be allegorized as the hoopoe. He is to men as the hoopoe is to birds, the one who knows the Name. This order of symbolism therefore depends upon auditory symbols in Arabic as the divine language of Islam. Nevertheless, it should be noted that this symbolism is not entirely unfamiliar in the West; it comes into Christianity from Islam during the period of the Crusades. The same associations sketched above are encrypted in the Christian stories of St Francis of Assisi conversing with the birds. There can be no doubt that such stories concern the appropriation of aspects of Sufi mendicantism into Latin Christendom through the Franciscan Order.

The Subterranean Waters

Have you not seen how Allah has sent down water from the sky
and has caused it to penetrate the earth as watersprings....
Qur'an 39:21

There is a further reason for the fame of the hoopoe bird, however, and from it is developed an order of symbolism that is of such prime importance to the Islamic religion that it can justifiably be described as indispensable. It concerns the ability of this bird to detect underground springs and other subterranean reserves of water. The hoopoe has a distinctive call but it is also renowned for finding water below parched lands, no doubt a marvelous and God-given skill in the estimation of desert dwellers such as the Arabs. To the Sufis, it is an extension of the same symbolism. The hoopoe bird knows the hidden Name of God because, in a parallel to the Sufic mode of exegesis of Scripture, it is able to divine what lies below superficial appearances. In this, subterranean waters are a metaphor for revelation.

The same metaphor is at the very heart of Islam inasmuch as Islam, historically, mythologically and indeed geographically, has the sanctuary at Mecca and its sacred well, the well of Zamzam, at its center. The foundation myth of Islam, which is re-enacted by the pilgrims to the Meccan sanctuary, concerns Hagar and Ishmael's search for water in the desert. Hagar runs seven times between two hills—noting the appearance of a similar motif in Attar's allegory, namely the journey through seven valleys—before she is granted divine assistance and the well of Zamzam is uncovered in Mecca valley. Similarly, the historical foundations of Islam concern the Prophet Muhammad's uncle rediscovering and excavating the well of Zamzam and then Muhammad's struggle to reclaim it for the God of Abraham. In all of this the well and its life-giving water represents revelation. The thirst of Hagar and Ishmael represents the thirst of man for salvation in the desert of the world. The story of Muhammad's grandfather re-opening the well is a quasi-historical account of the emergence of proto-Islamic currents surfacing among the Mecca tribes in the generations just prior to the Prophet. When we are told that 'Abd al Muttalib dug out the well just before Muhammad's time, we must appreciate that, while this might (or might not) be an historical fact at an archaeological level, it is, more importantly, a spiritual fact, namely that members of Muhammad's family had retapped the wellsprings of Abrahamic revelation and revived the Abrahamic cultus in Mecca.

The hoopoe bird, then, is analogous in these parallels to the Angel Jibreel who leads men to God's revelation and teaches men the Divine Names, facilitating communication between man and God, the seen and the Unseen. More specifically, though, it is the esoteric aspects of revelation with which the hoopoe is concerned because there are two dimensions to revelation, an inner and an outer, and it is the hidden waters that the bird can detect. The outer mode of revelation, therefore, has *rain*—the external waters—as its analogy. This is explicit in Quranic imagery and is, in any case, commonplace in all primal symbolism: God's revelation descends from the heavens as does rain from the sky. In contrast to the artesian waters, the waters of the rain are exoteric and represent the external character of revelation, although there is a further sense—namely where one leaves aside the contrast with artesian waters—in which rain represents the fullness of revelation. Of course, even subterranean water ultimately comes from rain, but it is divided off from surface water and so stands for that part of revelation which is hidden from the ignorant, or to use a more fitting term, the shallow.

All of this symbolism is once again the subject of the sacred precinct (*haram*) in Mecca. According to legend, Abraham and his son by Hagar, Ishmael, built—or re-established—the first shrine to the transcendent One God barely thirty metres from the opening of the well of Zamzam, namely the shrine of the Kaaba. There are many layers of association and legend superimposed upon one another, as well as extensive and often unintelligent 'renovations' that obscure and confuse the basic character of the Meccan sanctuary, and so the fundamental symbols involved in the Muslim Holy Places are no longer obvious and transparent, but we can still discern the basic idea. The well of Zamzam represents the hidden, esoteric waters, while the symbolism of the Kaaba is essentially to do with the analogy of revelation to rain. Renovations in modern times have disturbed the relationship between the well and the shrine in a fundamental way. First the well was pumped away from the Kaaba to make way for more pilgrims and then, in more recent times, it was prevented from flowing to the surface at all. It can now only be accessed through an underground grotto or through 'water coolers' placed around the Great Mosque that surrounds—or indeed intrudes into—the sanctuary. In earlier times the fundamental symbolism of the Kaaba was compromised when a roof and water-spouts were added. In its pristine state the Kaaba was open to the sky. We can only understand the most essential symbolism of the Kaaba if we appreciate that it was originally unroofed and exposed to sky and rain and that it is therefore the complement to the well of Zamzam, as vertical complements horizontal. The most important symbolism of the Kaaba is meteorological: sky, cloud, and rain are its primal elements. It was only so long as the Kaaba remained open to the rain—and the well remained open to the surface—that the most basic symbols of the sanctuary remained intact.

In extant accounts of the origins of the Kaaba its meteorological symbolism is often overlaid with later levels of pious rationale, although the original significances have not been entirely lost if we know how to 'decode' them. For example, in Shia traditions transmitted through the Imam Muhammad al-Baqir the sanctuary at Mecca was defined when God caused a pavilion from among the tents of Paradise to descend and the miraculous pavilion covered the exact area of the future sacred precinct. Here the 'pavilion' is nothing other than a symbol of the sky, the heavenly tabernacle, and the account describes how the sanctuary transposes an ouranic/celestial symbolism to terrestrial coordinates in a geomantic manner. The account of Ja'far as-Sadiq is more explicit. He relates that it was a White Cloud (*ghammah*), rather than a 'pavilion', that descended upon the sacred site and that the angel Jibreel ordered Adam to trace the area of the sanctuary at the limits of the shadow cast by the Cloud. As Uzdavinys observes in his article *The Sacred Institution of Haram in Mecca*, this Cloud is the 'spiritual realm of archetypes' that 'determines a terrestrial figure and becomes immanent to it.' But, to complete the symbolism, we must note that if the archetypal Cloud is pregnant with the Divine archetypes, then rain is revelation.

The Divine Throne

Say: Who is Lord of the seven heavens,
and Lord of the Tremendous Throne?
Qur'an 23:86

We reach a further abstraction of this symbolism when we encounter descriptions of the Kaaba as the earthly correlate to the Divine Throne. The Cloud symbolism is the most primal and Adamic. By contrast, the tabernacle symbolism is nomadic in character and Mosaic. The symbolism of the Divine Throne is architectural and, in prophetic terms, Solomonic. It is said that

the pilgrims circumambulate the Kaaba just as the angels circumambulate the Divine Throne. 'This is so,' as Uzdavinys says,
'because cosmologically the Kaaba is regarded as a reflection of
the archetypal Divine House' and is 'based on the metaphysical
premise ... that everything in the lower world is a projection
and an image of something in the higher noetic worlds ... a projection and an image of its spiritual archetype.' But this Throne
('arsh) symbolism is merely a concretizing of the earlier Tabernacle and Cloud symbolism. The sky is envisaged not as a pavilion
supported by four posts but as a Throne supported by four col-
umns from which God, now envisaged as King, dispenses His
commands. Needless to say, various meteorological symbols are
transposed into royal attributes; rain becomes the generosity
and wisdom of the King, thunder His judgements, shade His
Mercy, and so on.

The extrapolation of Throne symbolism to earlier orders of
symbols is, in fact, the purpose of the Quranic episode in which
the hoopoe bird first appears to Solomon. When the hoopoe
says, 'I have found out *a thing* that you apprehend not, and I
come to you from Sheba with sure tidings...' the main point of
his report is that he has found a people who worship the Sun
rather than Allah alone. Yet their application of Throne symbolism to cosmological realities is erroneous and diabolical and
needs to be corrected:

> Lo! I found a woman ruling over them, and she has been given
> abundance of all things, and hers is a mighty throne. I found her
> and her people worshipping the sun instead of Allah; and Satan
> makes their works fairseeming to them, and debars them from the
> way of Truth, so that they go not aright. So that they worship not
> Allah, Who brings forth the hidden in the heavens and the earth.

Sheba's 'mighty throne' entails false worship, and though it
yields abundance this only serves to disguise metaphysical error,
namely a form of paganism in which cosmological levels of
symbolism—in this case solar—have attained a spurious inde-
pendence from the transcendent Lord of All. The hoopoe, who
knows 'the hidden in the heavens and the earth' declares the
true God and the true symbolism:

Allah! there is no God except Him, the Lord of the Tremendous Throne!

In a story of strange twists that has confounded many readers, Solomon fools the Queen of Sheba with a fake throne and makes her mistake a paving of glass for a pool of water, thus demonstrating to her the difference between illusion and reality, appearance and truth, and at length causing her to submit to the true God.

> He said: Disguise her throne for her that we may see whether she will go aright or be of those not rightly guided. So, when she came, it was said [to her]: Is your throne like this? She said: [It is] as though it were the very one.

> It was said to her: Enter the hall. And when she saw it she deemed it a pool and bared her legs. [Solomon] said: Lo! it is a hall, made smooth, of glass. She said: My Lord! Lo! I have wronged myself, and I surrender with Solomon to Allah, the Lord of the Worlds.

Solar and Planetary Symbols

The very same rectification of solar symbolism is once again found in the Meccan cultus and in this instance serves to bring the Quranic story of Solomon and Sheba into focus. Once again, however, the so-called 'renovations' of the Holy Places in Mecca have done damage to the primordial symbols and the rites of the Hajj have lost much of their deeper significance, instead degenerating into merely commemorative superstitions. It is not only

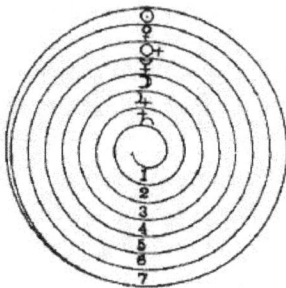

the relationship between the well of Zamzam and the Kaaba that has been disturbed, but also the relationship between those two symbols and the hills between which, according to the myth, Hagar scampered in her frantic search for water. The original arrangement had the well and the Kaaba situated in an extension of the natural depression between the two hills of Safa and Marwa. Today, both hills have been completely subsumed by the monolithic Great Mosque and so are enclosed indoors. One of the hills has been leveled to accommodate a roadway and, worse still, there are automated pavements—travellators—carrying pilgrims from hill to hill, or what remains of them. Moreover, even despite the fact that this section of the pilgrimage re-enacts mother Hagar running between the hills, it is now deemed improper for *women* to run any part of the circuit—only men are allowed to run in the designated 'green zone' in what was once the hollow between the hills. (And in a final move towards turning the Holy Places into little more than an Islamic theme park, the Saudi authorities are now planning to air-condition the whole structure!) Before these degradations the pilgrims moved from hilltop to hilltop seven times and resting at each vantage point looked down into the valley as Hagar did in her search for water. The pace of the pilgrims slowed to a walk as they climbed each hill and then accelerated into a bolt as they descended into the vale between.

This rite has an entirely solar symbolism. The two hills represent the two tropics defined by the furthermost limits of the Sun's movements along the horizon throughout its yearly course. As the Sun approaches the tropics each year its daily motion slows until it reaches the solstitial extreme where it seems to linger for several days. As it resumes its course towards the other tropic, however, it picks up pace, its increments along the horizon being most rapid at the time of the equinox which, in the Hajj rites, is represented by the low-lying so-called 'green zone' where (male) pilgrims break into a jog. The seven circuits represent the obedience of the seven ancient planets, with the Sun at their head, to this celestial ritual.

A complementary symbolism is found in the circumambulation of the Kaaba. While the circling pilgrims are no doubt emulating the archetypal movement of the angels about the Divine Throne, on a cosmological level they are also emulating the geocentric motion of the planets around the Earth. Again there are seven circuits, but in this case three of them are at a slow pace and four at a quicker pace since the extra-terrestrial planets move in slow orbits and the intra-terrestrial planets move fast by comparison. The rites of running between hills—the *Sai*—has a linear symbolism that concerns the motion of the heavenly bodies along the horizon and involves an order of astrological ideas such as we find in the ancient Book of Enoch. The rites of circumambulation—the *Tawaaf*—concern the astrological schema projected onto a plane as it is in the more familiar geo-centric astrological models still known today. The one, we might say, is a latitudinal symbolism, and the other longitudinal. Both are integral parts of the complete Hajj rites. The pilgrim must accomplish both journeys, from north to south (*Sai*) and around the center (*Tawaaf*).

Rectifications

These, like most aspects of the Meccan cultus, are pre-Islamic. Islam did not invent them, but merely rectified them, drawing them into a revivified Abrahamic mythology and installing the God of Abraham at their core. This is what Solomon does to Sheba and her pagan sun cult. She does not know a true throne from the illusion of one nor a true pool of water from a pave-ment of glass. Mighty though she is she must submit to Solomon and acknowledge He that the hoopoe proclaims as the 'Lord of the Tremendous Throne'. Islam restores a proper metaphysical perspective to the degenerating and decadent astral cults of the ancient Middle East. It brings the astral paganism of *jahiliyya* Mecca back into metaphysical alignment by reasserting the tran-scendent authority of the One God over all created symbols. It is significant that, on the face of it, there was nothing amiss with Sheba's religion: 'she has been given *abundance* of all things . . . and Satan makes their works fairseeming to them...' Solomon must expose such simulacra. It is the first tenet of Sufic psychology that the Self is hidden by the false ego, the *nafs*, an imposter who merely *seems* real. The quest is always to distinguish between what seems and what is. In a not dissimilar way, and also drawing upon a rehabilitated astral symbolism, the birds in

Attar's allegory must search the seven valleys for themselves, although Attar's symbolism is lunar, the thirty birds (*simurgh*) representing the thirty days of the lunar cycle.

And We have built above you seven strong heavens,
and have appointed a dazzling lamp, and have sent down
from the rainy clouds abundant water.
Qur'an 78:12–13

To further explore these associations we would need to enter a complex realm where symbolisms that seem antithetical are in fact interchangeable. At the very deepest levels various symbolisms converge. It would take us too far afield to explain every implication of this. But there are, all the same, a number of points that require elucidation in order to complete this present study and to bring it to a satisfactory conclusion that avoids confusions and misunderstandings.

Firstly, and in reference to the last point regarding Attar's symbolism being lunar, we must appreciate that in traditional cosmological thought Earth, Sun and Moon are three interchangeable poles, as the fact of an eclipse makes plain. Most importantly, Sun and Earth—heliocentric and geocentric viewpoints—are interchangeable depending upon one's position. The Kaaba, and the Divine Throne, might be either solar or terrestrial, or indeed, in another aspect, it might be lunar. Each might be axial. Another feature of the Hajj rites is ritual lapidation, the stoning of the pillars of Satan. This, alas, has also been degraded as recently as 2004 when the Saudi authorities demolished the *pillars* and replaced them with long *walls* instead, apparently for no better reason than that walls are easier to hit. In any case, the three traditional pillars represent Sun, Earth and Moon respectively—they must be pillars because the relevant symbolism here is axial—and stoning them (with seven pebbles) is once again a gesture of bringing astral symbols—debased by paganism—back into metaphysical alignment by way of the Abrahamic mythos.

Secondly, and as paradoxical as it might seem, the symbolism of water and *light* may be interchangeable in certain circumstances. If Hagar's search for water is analogous to the Sun's passage between the tropics then the well that opens up for her must be analogous to the Sun in its primal aspect and its water must be analogous to light. This is indeed the case. Zamzam, like other sacred wells, becomes a solar symbol. Conversely, the Sun is a *well of light.* Similarly rain and light are *both* symbols of revelation in the Qur'an and throughout other spiritual traditions.[1] The primal mythology active in this case is that celestial *light* is sown in the *clouds* where it condenses into *rain* which then falls upon the *Earth.* When we speak of the *descent* of the Divine we might more accurately speak in terms of an alchemical *condensation.* In fact, the four primal elements (of which the four-sided Kaaba bespeaks) are light, air, water and crystal (silica), all translucent substances, or rather the one translucent substance, at successively greater densities. Air is condensed light—this is important to all cloud symbolism—water is condensed air, and crystal (silica) is condensed water. This explains why the Qur'an allows sand (silica) to be used for ritual purifications instead of water in certain circumstances. The rites of purification in Islam (*wudu*) are actually accomplished with light (as water, or sand).

1. Revelations are 'light-giving' in the Qur'an. 'Their apostles came to them with veritable signs, with scriptures, and with the light-giving Book' (Qur'an 35:25).

The same interchangeability of water and light is found in ancient baptism cults of the Middle East such as the Hemaero- obaptists (Daybathers) and in some of the Dead Sea Scroll texts. Some of the miraculous qualities attributed to Zamzam water in Muslim superstition need to be understood in terms of the waters of the well representing the solar essence.

Thirdly, in a further complication, light and sound are inter- changeable. The symbolism of the hoopoe bird depends upon the Divine Name (the substance of revelation) being analogous to life-giving water, but in another layer of symbolism the lumi- nous principle is interchangeable with sound. In Biblical terms, God *said* 'Fiat Lux'. His speech and the primal Light are simulta- neous. A solar symbolism is introduced into Sufi recitals when, as is usual, the chant of the Name—Hu! Hu! Hu!—is accompa- nied by a simple drum-beat. Each beat of the drum represents a solar day, the solar pulse, the pulse or inhalation of day and night.

Finally, as a point of method, we must add here that there is often a danger in a strongly nominalistic tradition such as Islam to overlook a second level of symbols. Uzdavinys, for example, gives an account of the socio-historical aspects of the Meccan cultus and their metaphysical significances, but overlooks the cosmological and astrological order of symbols that stands and mediates between these two extremes. We ought not leap from a consideration of the socio-historical and tribal meanings of the Hajj rites (the terrestrial level) to an account of angels circu- mambulating the Divine Throne (the metaphysical level) with- out appreciating the essential astro-planetary symbols that are in play at the cosmological level. And, as we have seen, there is a separate level of meteorological symbols that are of prime importance also. The modern mentality expects neat divisions and simple equations such that a symbolizes b (and always and only symbolizes b). In fact, traditional symbolism is a complex algebra in which the multi-valance of symbols—in the service of paradox—is the first rule.

The Crescent & the Star

Notes on the *Hilal*
as a Symbol of Islam

The hornéd Moon with one bright star
Within the nether tip...
S. T. Coleridge,
Rhyme of the Ancient Mariner

Introduction

NO religion is born with all its symbols ready-made and fully developed. Rather, an order of symbolism is implicit from the beginning and, typically, is revealed or unfolded gradually over the course of time. It was some five hundred or so years after Christ before the crucifix became the acknowledged symbol of the Christian faith, and the Mogen David (Shield of David, Star of David or Seal of Solomon) has, surprisingly, only been a symbol of Judaism for the last two hundred years or so. In the case of Islam the crescent, or the crescent and a star, (hereafter called the *hilal*, the Arabic word for the lunar crescent of the New Moon) is the generally acknowledged symbol of the Islamic faith, yet it was not formally adopted into this role until quite late in the Ottoman Caliphate, which is to say in early modern times; it was not a part of Islam from the outset. Like Christianity and Judaism, Islam had no defining symbol in the beginning. No emblem accompanied the Islamic revelation. The Prophet Muhammad endorsed no symbol. Instead, the Muslim religion, like others, acquired its symbol in the course of history, and even then quite recently.

It is for this reason that many contemporary Muslim purists reject the use of the *hilal* and dismiss it as an 'innovation', a deviation from the pristine symbolessness of early Islam. They argue that since neither the Prophet, nor his family, nor his companions nor the early generations of Muslims knew of any such symbol, it can have no proper place in the pure and true form of the religion. So-called 'fundamentalists' have made it a matter of controversy. It is one of the issues around which the tensions between tradition and modernity among contemporary Muslims will sometimes coalesce. Most Muslims, and most non-Muslims, readily employ the *hilal* as the sign for Islam, and regard it as an established tradition, but there are so-called 'hardline' elements who shun it as part of their return to 'fundamentals' and to the 'true sunnah' (example) of the Prophet. At the same time, the fundamentalists—strange to relate—have taken heed of their counterparts in the Christian faith, who, in their mobilization against Islam, have attacked the symbol as a remnant of paganism and sure evidence that Islam is a sinister pagan moon cult disguised as a semitic monotheism. Not only is the *hilal* not sunnah but it is a symbol of the paganism that the Prophet and his Companions strove to eliminate and replace with pure worship.

This article will offer some brief notes on this matter, including some discussion on the significance of the symbol itself, and its origins, and of its relation, historic and intrinsic, to Islamic spirituality. While it appears to be a somewhat peripheral issue, it nevertheless raises some important matters and presents an opportunity to explore some things which, while they are little understood, are central to Islam.

Symbols and Trademarks

Firstly, let us admit that on this particular matter the hardline taken by the fundamentalists has some merit and is a case where they are in part right but largely for wrong reasons. Their opposition to the *hilal* is based on literalist, ossified readings of the sources of Islam and shows no appreciation of the quite orthodox ways in which living religions permutate beyond their initial forms; thus they are just as opposed to Sufism because the official formation of the Sufi brotherhoods occurred after the time of the Prophet and the Companions. But their stance does at least bring into question the whole procedure of allotting symbols to religions and draws attention to the fact that it is quite a modern thing to do. The idea that a religion requires a symbol as an identification marker is an outcome of the modern mentality where all things have such markers or 'trademarks' or 'brand logos'. It is modern bourgeois man that wants his ideologies packaged and brand marked. The *hilal* is not a traditional symbol of Islam, not a part its integral heraldry, rather it is the Muslim brand logo, its label in the modern ideological and spiritual 'market'. It became attached to Islam only when Islam needed a trademark, which is to say when it entered modernity and came under the sway of conditions of commodification. There is undoubtedly something irksome about the mentality that wants religions tagged like designer jeans. It is a manifestation of the same mentality that would tag Jewish citizens of modern states with the Star of David so as to dehumanize them, which is in fact how the Star of David became the 'brand logo' of Judaism. The opponents of the *hilal* are therefore right to regard the symbol as suspect, especially in so far as they draw attention to the degrading effects of such labels in general. Their anti-modern instincts are quite astute here. It is indeed confounding to every modern mind to hear that Islam is a religion that denies having a symbol. A religion, an ideology, without a logo? And no doubt they are right to say that when early Muslims still basked in the radiance of the Islamic theophany Islam was symbolless except for the plain black banner of the Prophet, itself symbolic of the black of the desert night.

The Politics of the Hilal

But, as in most cases, the fundamentalist position on this has little to do with a legitimate critique of modernity and much more to do with profane aspirations dressed up as religious piety. The modern rejection of the *hilal* by Muslims has more to do with politics and ethnic nationalism than it does with religion. We see this quite simply by looking at the flags of modern Muslim nations. The *hilal* features on many of them, but not all. In fact, it features on the flags of virtually all Muslim nation states, with the conspicuous exception of the Arab states. The Turkish flag features a white *hilal* design with crescent and five pointed star on a red background. The Pakistani flag has a similar design but with the Moon reclining on a green background. Malaysia has a yellow crescent and fourteen pointed star on a blue background. The flag of Turkmenistan has a waxing *hilal* with five five-pointed stars. The post-Soviet Muslim republics of Central Asian all have their own variations. But not the Arab states. The modern Arab states have not adopted the *hilal* as an indication of their heritage.

This is an anti-Turkish Arab-nationalist gesture. The Arabs associate the crescent with the Turks and with Turkish domination prior to the 20th century. It therefore does not feature as a symbol of Islam among the modern Arabs. Anti-crescent sentiment, manifest as religious puritanism, is an extension of this. If many Muslims throughout the world question the validity of the *hilal* it is because of the outreach of modern Arab Islam and especially the sanctioned Islam of the Arabian Holy Land, Wahhabism. The Wahhabi ideology is, in many ways, inseparable from Arab nationalism. On the authority of various *ahadith* (prophetic narratives) the Wahhabis draw a line after the first three generations of Muslims and declare that everything beyond that date is 'innovation'. This has the effect of anathematizing all of the distinct contributions of the Turkish, Persian and Indian peoples to the Islamic spiritual heritage and is a way of ethnically cleansing Islam. The Wahhabis accompanied the House of Saud to power in Arabia in the 1920s and, by means of oil wealth, have propagated Wahhabi ideology throughout the Muslim world and especially to newly formed and struggling Muslim nations. The assault on the *hilal* is as much motivated by this neo-Arabism and hostility to all things Turkish as it is by genuinely religious concerns. According to the opponents of the *hilal* the crescent and star was an ancient Byzantine symbol indicating the patronage of the goddess Artemis, and it was appropriated by the Turks when they took Constantinople. It is a pagan symbol, they say, recklessly acquired by the innovationist and heresy-loving Turks and it is precisely this of which the whole of modern post-Ottoman Islam must be purged.

Debate over the *hilal* therefore takes us to the heart of the plight and predicament of Islam in our times. The fundamentalism of which we have been speaking is a manifestation of the post-Ottoman era. When the decrepit Ottoman Empire, and the Caliphate, was dismantled the classical phase of Islamic civilization ended; the door was open for the assertion of Arab nationalism, and with this came the Wahhabism of the Saudis. These were catastrophic events, a terrible rupture of the traditional forms of Islam. In the recriminations after Ataturk and his followers took control of the Turkish state the Sufi brotherhoods were outlawed and the traditional Islam of the Turks was suppressed and brought under state supervision. The continuity of Turkoman Islam, unbroken since the Middle Ages, was sundered. At the same time the Wahhabis who were empowered by the demise of the Ottomans were viciously anti-Sufi and clung to the most horizontal of literalist understandings of Islam which they have subsequently propagated throughout the Ummah. By these events Islam entered a new depth of 'forgetfulness', something of its mystical perfume faded, it fell yet further into metaphysical decline and many of its secrets were lost.

The Meaning of the Hilal

This is the context in which we must consider the question of the validity of the *hilal*. Is it just a brand logo? Is it an ancient symbol of the Moon goddess that the Turks unwittingly imported into Islam? Or of the pre-Islamic Moon goddess of southern Arabia? Or does it have a symbolic value—a secret— that has now become obscure and unfamiliar to modern Muslims? Modern Turks will themselves tell you that the crescent and star is an ancient Turkoman symbol that their nomad ancestors carried across Central Asia and established as the banner of Islam upon adopting the faith. Is it then merely an ethnic symbol after all? Perhaps its association with Islam is accidental and it has no profound meaning? We must remember that modernity is a forgetting. The *hilal* emerged as the emblem of Islam from the tumult of the collapse of the Caliphate and thus from the religion's passage into modernity. Is it a symbol that survived from the closing phases of classical Islamic civilization but of which the significance has been forgotten?

There is certainly no consensus of opinion as to what the symbol actually means. This is why there are so many different variations used on flags and other Muslim insignia. The common elements are a crescent and a star, but there is no standard way to arrange them, except to say that the star is very often intruding into or is placed within the dark of the Moon's face. How does this represent Islam? What does it mean? The crescent alone is easy to explain: it represents the crescent moons by which the Muslim calendar is calculated, the *hilal* Moon. Sighting and calculating the *hilal* is a Muslim past time today and in previous centuries was a serious scientific endeavor. It is understandable and appropriate as an emblem of Islam, and indeed the crescent is often seen portrayed alone, as in the Red Crescent of the emergency humanitarian organization, the Muslim equivalent to the International Red Cross. But the crescent is usually accompanied by a star, and this creates difficulties. Which star? Which astronomical (or astrological) configuration is being depicted? And how does this appropriately represent Islam? It seems there are no clear answers to any of these questions.

There is, of course, a plethora of modern theories, but none are very convincing. Most are concerned to explain how a star can appear inside the dark of the Moon. Some argue that the star is actually the flash caused by a meteorite striking the Moon's surface in some prehistoric catastrophe, and that the *hilal* is a remnant memory of this event. Some will tell you that the design depicts a visitation by alien spacecraft that had flown across the face of the Moon in past centuries, which the poor simpletons of those days mistook for a divine sign. Others will explain it by atmospheric optical tricks and statistics about freak astronomical conditions. Others have calculated that there was a very bright conjunction of Moon and Jupiter near to the time that the Prophet Muhammad had his first vision of the Angel Gabriel and propose that the *hilal* depicts this event. Islam-haters will say that the *hilal* symbolizes the horns of the devil and Satan's star, Lucifer. Neo-pagans will say that it represents the Goddess and her Consort. It is very hard to find any sort of cogent and credible explanation among the current speculation from the current modes of inquiry. Historical research reports that in Ottoman usage the star was officially fixed as five-pointed in the late 1700s and was explained *post hoc* as representing the so-called 'five pillars of Islam'. But this only explains why the star is five-pointed, not why it is placed inside the dark of the Moon's face, and nor does it reveal what star it might be. What is this star of Islam?

The star is the riddle. The problem is to identify the star and then explain why it is placed inside the dark of the Moon. And then the problem becomes explaining what that has to do with Islamic spirituality. It would be easy to mount a full study of these questions. The symbolism of crescent and star is ancient—indeed crescent and star are both primordial symbols—and there is a profusion of theories that could be discussed along with ample archeological and historical examples on stone and metal, ancient coins and other media in times and places far afield. Given the limited scope of this present exposition, however, the most we can do here is make some general observations and attempt to point the reader towards the correct—which is to say traditional, substantial, integral—order of symbols and steer them away from empty modern speculation.

Several Hilals

Much of the uncertainty and confusion regarding the symbol arises from it being the combination or convergence of several very ancient but related symbols. Modern accounts often fail to distinguish between different symbols and collapse similar symbols into one. If we conduct a careful survey of *hilal* symbols through history we will notice that several distinct but similar symbols have converged in modern times, and this in large measure is what confuses things for us. We have only one symbol and are looking for a single answer to our questions when in fact the single symbol we know is an amalgamation of several historical precedents. The modern *hilal*, that is to say, is a synthetic symbol in which several earlier traditions of symbolism have been reduced into one design, largely as a consequence of the modern need for standardized logos as discussed above.

To be specific, the evidence requires that we distinguish between the crescent/star symbol and those symbols depicting a crescent and stylized sun. The crescent and star is one symbol and the crescent and sun is another, although they can look similar or even indistinguishable outside of their context since the sun can be and very often is, in some cultures, shown as a star. Sometimes the only difference between a star and the sun is the number of arms on the figure. Among the modern representations of the *hilal* only the flag of Malaysia still has a crescent and sunburst, but the earlier Ottoman designs were all Moon with sun, not Moon with star. This has been forgotten and with it an essential dimension of the *hilal* symbol. The modern designs have been standardized and the ambiguity between star and sun in earlier designs has been eliminated in favour of the star and at the sun's expense.

As well as this, there are several—or at least two—stars, or rather we should say planets, that regularly appear in ancient *hilal* designs and have behind them distinct traditions of symbolism, both of them distinct from the crescent/sun traditions. There are at least two quite distinct crescent/star traditions that have converged, and modern minds have forgotten both of

them as well as the distinction between them. These two stars
are Venus and Saturn. Sometimes the star that intrudes upon the
crescent is Venus, the evening/morning star, but there is
another quite distinct tradition in which it is Saturn, last of the
seven ancient planets. Again, visual codes such as the number of
arms on the star that would have made the distinction clear to
people in past ages have been entirely forgotten and along with
them a whole order of subtle understandings.

Despite this diversity of precedents, however, there is a single
motif that unites these different symbols into a single order of
symbolism and so there is some basis for the collapsing of them
into one symbol. The common factor is that in all cases the
Moon is in the act of devouring. The *hilal* depicts the Moon eating either star or sun. This is the whole key to the symbol. The
hilal is sometimes called 'crescent facing a star' but in fact the
crescent is devouring a star (if that star is not in fact the sun.) We
see this as an obvious fact the moment we put aside our modern
sophistications and see primordially, with the eyes of Adam, so
to speak. The modern mind can be ridiculously literal. People
will ask, 'How can a star appear across the face of the Moon? It is
impossible because no star could pass between the earth and its
satellite. The earth would burn up!' The star, of course, is within
the Moon, in its belly, having been consumed by the Moon. The
horns of the Moon are the jaws of its mouth. The crescent is a
mouth and the *hilal*—regardless of whether the star/sun is
entering or is within the shaded part of the Moon's face—shows
the Moon in the act of eating. It is sometimes shown eating the
Sun, sometimes Venus and sometimes Saturn.

We are therefore considering the symbolism of the lunar
dragon. To the primordial imagination the crescent Moon is like
the wide opened mouth of a dragon, and from time to time it
swallows either stars or planets by occultation, or the Sun itself
by eclipse. This is what the *hilal* is about at its most primal level. It
is a depiction of the lunar dragon and it is associated with everything suggested by the lunar dragon and its symbolism.

The Eclipse Hilal

In regard to the Moon eating the Sun we are, of course, talking about the mystery of eclipses. In an eclipse the dragon swallows the sun. This is why dragons are depicted as having fire in their belly and why they guard over a treasury of solar gold stashed in the darkness of the Earth. In mythology the chivalric and spiritual hero must conquer this dragon and recover the gold from the darkness.

All of this, and much else in many related myths and legends, refers—by cosmic metaphor—to the spiritual process of reconciling opposites that also has as its symbol 'the sun at midnight' (and conversely the stars at midday). This describes the state of spiritual realization where day or solar consciousness enters the darkest realms of sleep. This is the secret to the prophetic and the oracular. The recorded descriptions of the prophetic experiences of Muhammad are highly suggestive in this regard. In the state of revelation Muhammad bridged waking and sleeping; it is a state beyond both, a marriage and reconciliation of the most basic polarities in the human soul. The trance-state of the oracle at Delphi as well as the mode of the Abrahamic theophany in the biblical tradition are the same: 'And Abram fell into a deep sleep....' In historical terms, it is the solar *hilal* that the Turks carry with them from Central Asia. It is essentially shamanic.

As a coming together of opposites the eclipse, in Islam, also prefigures the Hour of Doom, the Day of Resurrection. In the hadith literature we find accounts in which the Prophet describes eclipses as forewarnings of the approaching Day. And in his Sunnah he offers special prayers at the eclipse. In an eclipse the night/day polarity is reversed. It forewarns of the Hour of Doom when the sun will rise in the west and set in the east and the whole of time will be reversed. In some Sufi groups, especially those with Turkish roots, we find a practice in which the *murid* (pupil) must replay their waking life backwards in their minds—like a film played in reverse—every night as they fall into sleep. This practice is directly related to the eclipse symbolism we are considering here. It is an extremely important symbolism in many traditions, Islam included.

One *hadith* of the Prophet is particularly illuminating regarding the metaphysical significance of eclipses:

> It is reported—(and Allah knows best)—that a solar eclipse occurred during the lifetime of the Prophet Muhammad—peace be upon him. He offered the eclipse prayer and afterwards his companions said to him, 'O Allah's Apostle! We saw you trying to grasp at something while standing at your place and then we saw you retreating.' The Prophet said, 'I was shown Paradise and wanted to have a bunch of fruit from it. Had I taken it, you would have eaten from it as long as the world remains.'

The Prophet could see into Paradise and indeed so real, so corporeal, was it to him he could have reached out and tasted its fruit. Paradise is no mere vision. It is supernal and super-real. The opposites that are reconciled here are spirit and body, *pneuma* and *soma*, world and God. In this *hadith* the Prophet has a glimpse of the resurrected state. The eclipse is like a doorway into that state, a realm beyond opposites.

The Venus Hilal

In regard to the Moon eating Venus it is emblematic of the Quranic theophany and of the conjunction Prophet/Angel or Muhammad/Jibreel, Muhammad being the Moon and the angel the evening star. At first consideration Venus seems the most likely of all candidates as the star of the *hilal* because she is often seen in the western sky accompanying a crescent Moon, and she is sometimes 'occluded' behind the Moon, in which case the *hilal* illustrates an occultation of Venus. The five-pointed star very often indicates the Venus *hilal* because the five points of the star illustrate the pentagram that Venus inscribes on her circuits of the sun as seen by the geocentric observer.

It is probably Venus to which the mysterious early Meccan *surah* called 'The Night Visitor' in the Qur'an refers:

> Consider the heavens and the night visitor.
> What could make you understand what comes in the night?
> It is the star that pierces through darkness
> For no human being has ever been left unguarded.

The key Arabic term here, used metaphorically, is *at-tariq* with the root meaning 'to knock' and the idea is that a certain star 'knocks' at a door like a visitor in the night. It is sometimes argued in modern commentaries that this must refer to Venus as the morning star since it appears, or visits, (very late) in the night, and is thus a 'night visitor'. Maybe so, but the next few lines of the *surah* make it clear that it is more the evening star knocking on the gates of night to which the Qur'an is alluding in the first instance. It is the evening Venus that loiters above the dusk horizon like someone standing at the doors of night waiting to be let in. The *surah* continues:

> Let man then observe out of what he has been created: he has been created from the seminal fluid issuing from between the loins (of man) and the pelvis (of women)...

Venus is also, of course, the planet of love and rules over the union of man and woman, the sexual and procreative. It is Venus, the evening star, that guards over the conjunction of the opposites man and woman, night and day and the entry of the phallic sun into the womb of darkness. The core significance of the Venus *hilal*, therefore, is that where Venus is shown inside the Moon then the Moon has been impregnated. The short, ecstatic Meccan *surat* that announced the Islamic theophany often refer to the mysteries of embryonic fertilization. This reflects the spiritual fertilization of the Prophet's heart whereby the seed of the Qur'an—the whole Qur'an in principle—was implanted in his bosom by the Angel Jibreel on the Night of Power. We must remember that the Prophet in Islam has the parallel function to the Virgin Mary in Christianity, his unletteredness being exactly parallel to her virginity. Thus, too, the Qur'an—placed in the Prophet's heart—is parallel to Christ, placed in the Virgin's womb. At the Christian annunciation the Virgin is impregnated with the Logos, the Word. So too Prophet Muhammad on the Night of Power. This is what the Venus *hilal* represents, the germ of revelation, the pregnant theophany.

The Saturn Hilal

Finally, in regard the Moon devouring Saturn, modern readers need to be completely reacquainted with the symbolism of this planet before they are likely to understand the Saturn *hilal*. The discovery of the modern extra-Saturnian planets and hence the modern notion of the expanded 'solar system' has obscured the entire symbolism of Saturn as the last of the visible planets and the ring-pass-not of the traditional seven planet *kosmos*. Saturn is, in traditional understandings, the cosmic chronometer, the time-keeper, Father Time. But also, as Chronos, he is lord of the Golden Age. The importance of this symbolism in the Semitic traditions is signaled by Saturn being the lord of the sabbath, which in Judaism is on Saturn's day—day of rest. In Islam, a related symbolism concerns the famous 'Black Stone of Mecca' which is Saturnian in its symbolism—pilgrims touch or kiss it at the end of the seven (planetary) rounds of the Kaaba—and signifies both center-point and end-point. The Moon is the fastest moving of the traditional planets and Saturn is the slowest with a symmetry between the Moon's period of twenty-eight days and the twenty-eight year period of Saturn. The Saturn *hilal*, therefore, signifies 'First and Last', the first planet and the last. In an Islamic context it is equivalent to the Alpha and Omega used in Christianity. In terms of the symbolism of the lunar dragon it shows the dragon eating its own tail, the *ouroboros*, in turn symbolic of the cosmic cycle between Creation and the Day of Resurrection. Saturn is the planet of doom and death but also of the Golden Age and Paradise. The Saturn *hilal* carries the meaning: the end is a return to the beginning.

The communist development of the *hilal*, the 'Hammer and Sickle'—industry and agriculture—is directly related to this particular order of symbolism. Crescent = Sickle = Agriculture. The star of the *hilal* is the spark from the blow of the hammer on an anvil—industry. In terms of the symbolism with which we are concerned here it indicates the Ferric or Iron Age, the Kali Yuga, end of a cosmic cycle.

Conclusion

Although Islam is the most recent of the great world religions it claims to be both new and primordial, first and last. Its stark monotheism is both a culmination and a return. It is not surprising that, however it may have happened, Islam arrived in modern times bearing a primordial symbol. The symbolism of the cross is well documented, not least in René Guénon's work of that name. The symbolism of the Star of David or Seal of Solomon is equally well-known and understood. The symbolism of the *hilal* is a mystery by comparison. The whole modern tendency to label religions with convenient brand-marks is suspect, and there is some validity in resisting it, but at the same time anti-*hilal* sentiment in contemporary Islam is largely politically motivated and is certainly based on ignorance of what the crescent and star might mean in a broader frame of reference. We must remember that the historical religions, Islam included, are themselves a forgetting. The historical religions are built over wellsprings of a vastly more ancient symbolism that most religionists barely suspect, a fundamental order of symbols of which the ill-named 'fundamentalist' has no knowledge at all. It is hoped that this present article might stimulate further consideration of these matters.

Numbers & Letters

Modern and Traditional Perspectives on some Mysteries in the Qur'an as a Symbol of Islam

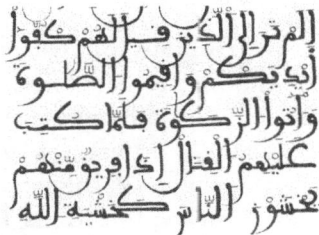

Allah sends astray whom He will
Qur'an 74:31

AMONG the idle and metaphysically illiterate chatter that is characteristic of our time are various attempts to demonstrate the truth of religion by scientific means. The grossest manifestations of this are such enterprises as archeological expeditions to eastern Turkey in search of Noah's Ark, but there are more subtle and ingenious examples, such as attempts to marry Biblical prophecy with the DNA code. In the same category are attempts to demonstrate the praeter-human source of various scriptures through numerical, statistical and computer analysis. In an age where all paradigms of intelligence exclude the possibility of revelation and so-called 'higher critics' untiringly expose the all-too-human events that have often been a background to the settling of textual canons, believers toil to show that their sacred texts are sacred in origin and supernatural in authorship. In both

Judaism and Christianity there has been a proliferation of pseudo-kabbalistic cryptography aimed not at exposing the metaphysical depths of the text but at demonstrating a type of scientific validity, something to convince the skeptics on their own ground. Behind these attempts is the assumption that divine truths can be quantified and demonstrated to the satisfaction of the scientist and statistician. Muslims have subjected the Qur'an to this type of spurious development. Widely-read modern commentaries on the Qur'an casually map the whole of Darwin's theory of evolution in the Divine utterances, miraculously foreseen in the seventh century Arabian text, while others resort to numerical exposition and complex computer analyses to demonstrate the text's miraculous foundations. The miraculous nature of the Qur'an is a matter of orthodox Muslim dogma; many modern exegetes, and much popular understanding, can only comprehend this claim by criteria set by profane science.

A particularly enduring form of this approach to the holy book of Islam has been the codifications investigated by Dr Rashid Khalifa, Imam of the Mosque of Tucson Arizona, in the 1970s and more recently promoted by scores of websites. Dr Khalifa claims that he discovered a numerical key based on the number nineteen, a mystery mentioned in the Seventy-Fourth *Surah* of the Qur'an. This was in 1974, which fact he took as immediately significant since the numeric for the year (albeit in the Gregorian calendar) is the combination of 19 and 74, the mystery and the *surah*. In Qur'an 74:30 we encounter the short, cryptic *aya*: 'Above it are nineteen.' And in the next *aya* we are told '[this] number have We made to be a stumbling-block for those who disbelieve....'

The full context is as follows:

27.—Ah, what will convey unto thee what that burning is!—
28. It leaveth naught; it spareth naught
29. It shrivelleth the man.
30. Above it are nineteen.
31. We have appointed only angels to be wardens of the Fire, and their number have We made to be a stumbling-block for those who disbelieve; that those to whom the Scripture hath been

given may have certainty, and that believers may increase in faith; and that those to whom the Scripture hath been given and believers may not doubt; and that those in whose hearts there is disease, and disbelievers, may say: What meaneth Allah by this similitude? Thus Allah sendeth astray whom He will, and whom He will He guideth. None knoweth the hosts of thy Lord save Him. This is naught else than a Reminder unto mortals.

From this Dr Khalifa and those who have followed him have extracted the 'Nineteen Code' as 'scientific' proof of the Quranic miracle. The entire text, according to this theory, is permeated with patterns of nineteen that no mortal mind could possibly have composed. The Qur'an, Dr Khalifa declared, is the ONLY scripture in existence with a built-in physical, examinable, and indisputable proof that it is God's message to the world.

In recent times the 'Nineteen Code' has swarmed about the Internet among Muslim proselytizers and apologists as 'proof' that the Holy Qur'an is not—cannot be—the work of human hands. It is inconceivable that a human writer could have hidden such a code in the text.

The Nineteen Code begins its examination with the Quranic exordium: *Bismallahir Rahmaanir Rahiim* (In the Name of God, the Beneficent, the Merciful) which is constituted of nineteen letters in the Arabic. Next, each word of this formula is found in the Qur'an a number of times that is some multiple of nineteen. The first component, the word *Ism* (Name) occurs nineteen times in the Qur'an. The second word, *Allah*, occurs 2,698 times, or 19 x 142. And so on. Moreover, the Qur'an consists of 114 *surat*, and 114 = 19 x 6. The formula *Bismallahir Rahmaanir Rahiim*, though, is only found at the beginning of 113 *surat*, one short of a multiple of nineteen: it is missing from *Surah* nine. This 'missing' *bismillah* is to be found in verse 30 of *Surah* 27 (noting that 30 + 27 = 57 = 19 x 3), the only case where the formula is found in the body of a *surah*. So there are, in total, 114 *bismillahs* in the Qur'an (114 = 19 x 6) and *surah* 27 is no less than nineteen *surat* from *surah* 9. To find the missing *bismillah*, that is, you count nineteen *surat* from the *surah* from which it is missing. Going further, in *Surah* Twenty-Seven, the number of words from the opening *bismillah* to where it appears in the body of the text at verse 30 is 342 words, or 19 x 18. Added to this, the first *surah* revealed, *Surah* Ninety-Six, consists of nineteen verses. The first section revealed (96:1–5) consists of nineteen words. These nineteen words consist of 76 letters or 19 x 4. And so the analysis continues, adding further and further examples of the miraculous 'Nineteen Code'.

This Code, according to its proponents, settles forever not only the mystery of 74:27–31 but a number of other perplexing aspects of the Quranic revelation. It explains, for example, why the *surat* have been organized as they have. This has been a matter of curiosity to readers of the Qur'an through the centuries, and a matter of some consternation to Western readers in modern times who cannot understand why the revelations to Prophet Muhammad have not been arranged in chronological order as seems natural to the Western mind. Instead, the *surat* seem to be organized somewhat haphazardly, roughly in order of length, longest to shortest. The Nineteen Code, it is said, shows that the canonical arrangement is not haphazard but integral. *Surat* have been shifted to encode patterns of nineteen throughout.

Furthermore, there is the well-known mystery of the so-called Abbreviated Letters, which letters appear without explanation at the beginning of certain *surat*. The meaning and significance of these letters has defied clarification throughout the centuries. It is agreed that they are an essential part of the complete text, but there is no agreement as to their significance. According to the 'Nineteen Code' the purpose of these letters is to manipulate certain letter frequencies and other statistical factors to ensure that all calculations add to nineteen or a multiple of nineteen. These otherwise cryptic letters are thus an important part of the numerical code that runs throughout the text.

It will be noticed that this theory defies the natural and contextual identity of the mysterious 'nineteen'. In context the number plainly refers to a number of angels set as wardens over the Hellfire. Challenged on this point the proponents of the theory resort to explaining that God saw fit to 'provide the earlier generations with a temporary interpretation for the number

19 of verse 30,' saving the 'true' interpretation for the modern era.

The most effective refutations of the Code have come from skeptics who have demonstrated that many texts—Herman Melville's *Moby Dick*—also contains the same or similarly dazzling patterns of nineteen, if one cares to search for them, and that it is not a unique feature of the Qur'an. The structures of language generate many patterns that might make those who notice them imagine such patterns are magical or miraculous.

It is not the purpose of this article to offer fresh refutations, but rather to bring some clarity to the clouded issue of these particularly arcane features of the Qur'an. The Qur'an, The Clear Guide, is certainly a cryptic text in places, and when it is it intends to be: 'What meaneth Allah by this similitude?' The Nineteen Code offers a solution to not only the mysterious 'Over it are nineteen' but to other mysteries, and indeed to the mystery of the very nature of the Qur'an itself, but it does so in ignorance of all traditional and metaphysical understandings. Dr Khalifa's 'discovery', it is said, was according to a Divine Plan and so, in fact, a further unfolding to the Quranic revelation. The true Quranic miracle had been hidden for centuries, forestalled until the age of unbelief and its antidote, computer frequency analysis. Rather than restoring any metaphysical perspicuity to these aspects of the Qur'an the Nineteen Code brings pseudo-science posing as an extension of revelation. The character of the theory is best exposed by restating the traditional doctrine and it is best refuted by giving modern readers a clear presentation of certain key facts about the Qur'an that have long been forgotten.

The Quranic mysteries encountered here—the cryptic reference to the number nineteen in *Surah* Seventy-Four, the Abbreviated Letters, and the sequence of the *surat*—can be clarified by one realization: the Qur'an is a complete recapitulation, in microcosm, of the entire cycle of existence. But, as in other respects, it is this at the level of structure, not narrative. There is an organizing principle in the Qur'an about which modern 'textual critics' have not the vaguest understanding. Yet it was the first duty of those who brought the Qur'an into its received form (*mushaf*); the principle that the whole must reflect the inner principles of manifest time. The Bible does the same but in the narrative mode. The Christian Bible begins at creation and ends with the apocalypse. Thus, in some editions, we find an Alpha printed on the front of the codex and an Omega printed on the back.

It is a matter of cardinal importance to realize that the Qur'an is organized according to the same design but not in the same style or at the same level. Instead, the organization of the longer *surat* at the beginning and the shorter *surat* at the end reflects the increasing pace of cyclic time and hence the rushing approach of the Hour of Doom. In traditional understandings of the passing of the Ages, let us recall, the Golden Age was of the longest and the Silver Age only a fraction of its duration, while the Bronze Age was only a fraction of that, and the Iron Age only a fraction of that. In cyclic terms the dissolution comes rapidly. When the center folds the periphery is dragged into an ever-hastening whirlpool, each stage more rapid, and shorter, than the last. This is why the message of the Qur'an is urgent and dire. It is not just that the Hour of Doom is near in time, but time itself is hastening. Temporal conditions are changing. Moments are getting shorter. The downward curve is not an even slope. The arrangement of the *surat* of the Qur'an is an expression of this doctrine. The opening revelation *Surah al-Fatihah* is Adamic. At the closing of the Book *Surah al-Ikhlas* is the *surah* that corresponds to and encapsulates the historical Islamic, final, revelation. The concluding two *surat* are 'seals' over the whole—in the End refuge is only with Allah.

This dimension of the Qur'an—it is true also of the Torah in Hebrew but not of the Christian Bible—extends to the very letters of the text. This is the key to understanding the so-called Abbreviated Letters. The pertinent fact to take into account is that there are fourteen such letters given privilege by being uttered in preface to certain *surat*, which is to say fourteen letters from an alphabet of twenty-eight. The symbolism is lunar. Each letter corresponds to a day of the lunar cycle of twenty-eight days. In fourteen of these days the Moon waxes and in fourteen it wanes, light and dark, revealed and concealed. This is such fundamental Islamic symbolism that it is a measure of the profanation of modern Islam that it even needs to be restated here. Every letter of the Holy Qur'an represents a day of a lunar phase, and the entire text represents the complete duration of time, the exhaustion of all lunar cycles.

The number nineteen has a lunar meaning that is an integral part of this same design. It is indeed the number of the wardens set over the Hellfire, and as 17:31 suggests it is a key to certain important things. 'None knoweth the hosts of thy Lord save Him' but the number nineteen serves as 'a Reminder unto mortals', a human approximation. The proponents of the Nineteen Code are not mistaken in detecting a numerical device here—it is signaled in the plain reading of the text—but it has another meaning to the one they imagine. The number refers to the Metonic cycle, the nineteen year cycle of soli-lunar eclipse.

It is not necessary to describe the details of this cycle or to explore the significance of every aspect of this matter; it is enough to know that, in a world ruled by a lunar calendar, the slippage of time on a cosmic scale is known and measured by the inexactitude of the Metonic cycle, just as it is known and measured by the retrograde procession of the equinoxes in a solar calendar. The Jewish lunar calendar is adjusted by adding intercalary time each nineteen years to prevent this phenomenon. The Muslim calendar is a 'rectification' of the Jewish calendar to its pristine (Abrahamic) state, without intercalation. The 'guardianship' of the 'nineteen' needs to be understood in these terms. Amongst other things, they guard against intercalation. The pure Islamic calendar allows the slip of time. The Metonic cycle of nineteen solar years, the discovery of which is attributed to the ancient Greeks—a calculation of when the occurrence of eclipses returns to the same days of the solar year—is but a near approximation; in fact it is slightly short of the true conjunction of solar and lunar cycles. The Qur'an alludes to this. It offers the number nineteen, but this is only a 'reminder' to man. God alone knows the exact number.

This, we might say, is exactly why the *surat* of the Qur'an are not all of equal length, because although each letter is a lunar day, there is an inexorable and unavoidable diminishing and decline as time moves on. We can understand this better by comparing it to an example from the ancient Greek tradition. In Plato's *Republic* Socrates is asked why his Ideal State—of which 'there is a model writ in heaven'—will not, despite being the very best of polities, endure forever? He answers with a cryptic number, usually known as the 'nuptial number'. It is a mathematical, astronomic key to the question: why does the world wind down, why is there cyclic decline, why will the center not hold? It is generally agreed that this 'nuptial number' has to do with the calculation of the Great Year and the measures of how copy and model gradually drift apart. It is the measure of imperfection in the *kosmos*.

The number nineteen—the Metonic number—serves this same function in the Quranic lunar cosmology. The 'wardens of the Fire' are the nineteen years of the Meton cycle. If we understand the Meton cycle aright then it 'will convey unto thee what that burning is!' We will understand why time must decline and culminate in the cleansing fire that 'leaveth nought and spareth naught.' As for man, the apocalyptic fire 'shrivelleth' him—an image of how he is gradually diminished through the period of the whole cycle, for it is a cycle in the soul of man and not just in the heavens. What is the Fire? Ah, says the Qur'an, if only We could explain what it is! The clue is that We have set nineteen over it, but this is only a reminder for men, not the true number. The 'Fire' is that 'gap' between the ideal and the actual. The whole of creation must eventually succumb to the fact that it is not the Real.

These matters are all germane to the very reason why the full disclosure of the Qur'an—as a last warning, a last revelation—had finally become necessary in the affairs of men. Islam explains these things in terms of a pure lunar reckoning. We need only engage in some deep and thoughtful reading of the text with this in mind to shed light upon some of the Qur'an's more opaque features and passages. This restores a traditional symbolism and puts the wild speculations of Muslims in our own times into a sober perspective.

Glass & Stone, Radiance & Solidity

A Public Lecture at Sacred Heart Cathedral

In beginning to construct the body of All,
God made it of fire and earth ... radiance and solidity.
Plato, *Timaeus*, 31b

A cathedral is, by definition, a medieval creation, and so a modern cathedral is, some might say, an anachronism. A cathedral is the product of a specific world-view and, more than other buildings, a specific cosmology, and it is a world-view and a cosmology to which the modern world no longer subscribes. I look at Sacred Heart as a medievalist. When I came to Bendigo, as a medievalist I was drawn immediately to the cathedral. What is a gothic institution such as this doing so far in time and space from the European Middle Ages that first gave birth to such buildings? But of course Bendigo, while blessed to have a cathedral of this stature, is not peculiar in this regard. These medieval buildings—a specific inspiration of a specific combination of factors in the European Middle Ages—are found throughout the New World, in cities both major and middle sized. The key to the longevity of this particular architectural mode, this architectural style, is no doubt the institution of the Church herself. It was the Church that carried her traditions and values from the Old World—from Europe—to the New, and the Church herself is, of course, an institution that actually extends beyond the Middle Ages, having her origins in the ancient world. The modern Church still builds cathedrals in the medieval style because, for the modern Church, the medieval style is, quite rightly, emblematic of one of the richest and most fertile periods in the Church's own history, in the history of Christians as a whole, and indeed in the whole spiritual history of the human race. A cathedral such as Sacred Heart is an emblem—we might almost say, only an emblem because it is in truth a very modest building compared to its European prototypes—an emblem of a particular configuration of ideas that came together in the European Middle Ages—in the Christian Middle Ages. By any account the great cathedrals of the Christian Middle Ages represent the zenith of medieval Christian spirituality, a spirituality preserved in monuments of glass and stone. The modern Church continues to build cathedrals in this style not out of nostalgia or a lack of fresh ideas but because this style marks an extraordinary high-point, a key moment, in the life of the Christian faith, and the Catholic faith especially. This style of cathedral points to a distant epoch of history, and in the first instance at least it serves to remind us of the spiritual achievements of that age. That we find such a building in such a place as Bendigo Australia—which is a long way from anywhere—serves to remind us that these achievements, by their very nature, transcend that age. It is among the duties of the Church to preserve and communicate to new generations the achievements, the discoveries, of her history. A cathedral is a book. A cathedral is a message from the distant past. It invites us to think again about the Middle Ages, about the Christian Middle Ages, about a great era of Christian faith, of a great era of both action and contemplation. I am not a student of local history. I don't know what was in the minds of those who built this building, Sacred Heart, but the effect of their actions has been to place in Bendigo—displaced by thousands of miles, by a hemisphere—an emblem of the great Christian Middle Ages. I often bring students here to see the cathedral. Students of medieval studies. They are indeed fortunate to have a building like this to connect to their studies. One of the profound things that happens when students visit the building is the realization that it is *real*, that it is a Church, a place of prayer, that it houses a living tradition. Students come here and no doubt they expect the building to be as it is to them, like a museum, an old object of study, something from the distant past. They come here wanting to know more about pointed arches or the symbolism of the transept. But then, inside, as they come in with their pen and paper and their analytical eyes, they often encounter people praying, people at prayer. This makes a profound difference. This is the moment for them to consider the nature of the Church, a continuous living entity from the Middle Ages to today, or rather from the ancient world, through the Middle Ages, to today, continuous and unbroken. The building is impressive, but gothic architecture is, after all, only one episode in the history of the Church. My students study many aspects of medieval Christianity, and the great Cathedrals, their design, their history, their meaning, their symbols, their iconography, but everything is placed in a new perspective when students—many or even most of whom are not Catholics or even Christians as such—realize that a building like this is, in fact, a living sanctuary. It is easy for medievalists—historians generally perhaps—to develop a museum mentality. As a medievalist, when you enter here, a building like this, it is easy to be absorbed by its aura of the past. To be transported back in time. But this is a living tradition. This is a living building. A living sanctuary. The Church perpetuates the spirituality that, in the 12th and especially the 13th century, in the Latin West, found remarkable, astounding expression in amazing constructions of glass and stone. Again, I don't know what was in the minds of those who first conceived of this building, but the effect of their actions has been to place in our community a living connection with a fantastically rich spiritual tradition extending from the Middle Ages to the present day.

It is important, I think, to see a building such as this in this sort of light. Indeed, cathedrals everywhere—especially cathedrals in the New World—should be seen in relation to what it is they connect with in the Old World—that extraordinary conjunction of factors, of profound ideas, in the Middle Ages. I want you to consider Bendigo for a moment. Think of it before the arrival of white settlers. Think of this site. This hill. This gully. Now think of the European settlement that has been imposed upon this land. Think of how Europeans came, and with them they brought, quite naturally, the things of Europe. They built European buildings. They transplanted European institutions from Europe to the new land. When you look around Bendigo, Sacred Heart is the single most conspicuous transplantation of a European building, a European institution to this new land. Now that the spire is complete the cathedral takes on a geo-physical status. It is a landmark in the full sense. Not just a building. A landmark. As you approach Bendigo from whatever direction the spire says, 'European occupation ahead!' But it is important to realize that the spire proclaims the spiritual legacy of European civilization and not some other aspect. There are many ugly marks of European occupation. There are highways and polluted rivers. There are the mines which remind us that greed and opportunity—not spiritual aspiration—brought Europeans here in the first place. In major cities, in Melbourne and Sydney and Brisbane, in the cities of Europe and North America, everywhere, office buildings and skyscrapers now dwarf the great cathedrals. There are spectacular instances of this in cities like New York. Big cathedrals humbled by massive skyscrapers built on the block next door. Or cathedrals squeezed in between two massive office towers of a hundred storeys or more. Huge, steel and glass monuments to money and prosperity are now the landmarks of those cities and announce the more ominous aspects of European occupation of the New World. The skyscraper is the emblem of heroic capitalism, a secular, utilitarian world-view. As a medievalist I like living in a city where the cathedral is still the tallest building in town. It is like a proclamation of the primacy of spiritual over secular values, one of the noble characteristics of the Christian Middle Ages. The Middle Ages were a period in which it seemed only right and natural that affairs of the spirit—matters having to do with God, with salvation—came before all other things. The spire of a cathedral says that. In its singular pointedness, in its vertical thrust heavenwards, in its domination of the medieval landscape, it proclaims the primacy of things of the spirit. The act of building a building like Sacred Heart, far from that medieval landscape, insists on perpetuating that ideal—the primacy of spiritual things. Sacred Heart functions in the Bendigo landscape in a manner very similar to that of a cathedral in a medieval town. You can see it from all directions, the tallest object, the most conspicuous feature of the landscape. If you get a view of Sacred Heart from up on One Tree Hill or from the lookout in the Whipstick Forest, or from any similar vantage point, the spire is like a finger pointing upwards to the sky. Its singleness is impressive. It represents spiritual aspiration. To one looking from a distance, or approaching, it says, 'There are people here, there is a community here, who aspire to spiritual things.' We no longer have the same spatial view of Heaven that places it in the skies above. Heaven, God himself, are more abstract things for modern man. Our cosmology has changed. The heavenward thrust of the spire of a cathedral might more naturally remind us of rockets and escape velocity. The sky isn't a symbol of Heaven for us as it was for medieval man. But we understand the symbol nevertheless. And we still build buildings—cathedrals—that embody that symbolism.

In a similar but more profound and subtle manner our understandings of other aspects of our world have changed in fundamental ways. I mentioned skyscrapers, massive towers of steel and glass. A steel frame and a glass tower. Such buildings were made for cities in the north of Europe, where winters are long and pollution and other factors produce an environment of relatively poor light. Glass towers are made for such light-starved environments. They are ridiculously hot and difficult to keep cool in a light-filled environment like Australia. A cathedral is a creation made of stone and glass. A skyscraper is out of place in an environment such as ours. It is an inappropriate architecture. We might think that a cathedral is inappropriate in the same way, so far from its natural environment, the environment for which it was originally tailored. But a cathedral isn't just a building. It's an alternative theory of the universe. I wrote my doctoral thesis on a work by the ancient Greek philosopher, Plato. A work called the *Timaeus*. In the *Timaeus* Plato describes a universe constructed of geometrical relations. There are modern physicists who are very interested in his ideas, because Plato insists that matter is not the fundamental substance of the universe, but rather geometrical, mathematical relations conceived and maintained by some great Intelligence. At the beginning of the *Timaeus* we are also told that the universe is made of two fundamental principles; radiance and solidity. There is light— which in itself is an invisible thing—and there are solid objects that light shines upon. Radiance and solidity. Plato describes how immortal ideas—archetypes—are made manifest through the mysterious agency of light. It was this passage on radiance and solidity that St Augustine took up in his work *City of God* and compared to the teachings of the Book of Genesis. Throughout the Middle Ages this work by Plato—preserved in a Latin translation—was regarded as a parallel text to Genesis and provided one of the cornerstones of the medieval world-view. This same text was used by the writer of the Dionysian corpus—the *Mystical Theology* and other works—that had a profound influence in the Middle Ages. It was these ideas—ancient ideas appropriated by the Christian tradition—that Abbot Suger brought together in his chapel of St Denis, the first flowering of the gothic style. The gothic style is about the interplay between radiance and solidity. It is about the creative dialogue between these two fundamental things. You don't need a Ph.D. in ancient

Greek philosophy to realize this though. It is self-evident in the buildings themselves. If you consider the style of architecture that came before the gothic style—the so-called Romanesque style—the Romanesque building was characteristically heavy, squat, fortress-like, a defense. It had massive stone walls and tiny windows. A single material dominated—stone. But the gothic style is about two materials, stone and glass. Compared to the Romanesque buildings that came before them the gothic buildings of the high Middle Ages are characteristically light, slender, tall, elegant and there is nothing of the military fortress or the castle about them. They are made of two materials—stone and glass. The walls of the Romanesque building grow thinner. The stone is handled as if it was a soft, malleable, plastic material. It is carved and twisted and treated as if it was as soft as wood, compared to the massive blocks, the stoic sense of gravity, in the Romanesque style. The small windows of the Romanesque now open up and large sections of the walls now become glass. By the end of the Middle Ages some of the stranger cathedrals were almost like glasshouses. A cathedral is about light. It is about the way light shines through glass but reflects upon stone. The construction of the cathedrals was a great engineering feat, but it was also a profound metaphysical statement. A cathedral is about light, light conceived as a spiritual thing, the vehicle of creation, the vehicle of divine manifestation. In glass and stone, areas of glass, areas of stone, there is an interplay, a dialogue, a dialectic, a creative tension, between light and dark, that which is transparent and that which is solid, between radiance and solidity, in a gothic cathedral. The great rose windows of Notre Dame; they are not for purely aesthetic purposes. They were not made just to look nice. Their real significance is lost on modern viewers. They are really about the way in which God used, uses light to manifest and sustain His creation. They are a commentary on the scripture: *And God said let there be light and there was light.* In the writings of St Dionysius—following Plato, endorsed by Augustine—we are told:

Every creature, visible or invisible, is a light brought into being by the father of lights. This stone is a light to me. This stone becomes a light to me.... For I begin to think whence this stone was invested with its properties, and soon I am led through all things to that cause of all things, which endows them with place and order, with number, species and kind, goodness and beauty and essence....

The relationship between glass and stone in a building such as this is not an accident, nor even a convenience of engineering. It is a spiritual doctrine—a spiritual understanding of creation, of the universe, embodied in a building. The whole building should be understood in these terms. There are two principles, two forces: glass, stone; light, dark; radiance, solidity. Eternity. Creation. Also light and heavy, levity, gravity. Stone is a heavy material. Glass is a light material. In the building some materials fall earthward, while others are liberated from the earth. The Romanesque order is squat and square, rectilinear. A gothic building is above all marked by its *vertical* nature. Up and down are symbolic directions in a gothic building in a way they are not in a Romanesque building.

One morning I came to Sacred Heart—just after sunrise—and I walked around just watching the beams of the rising sun on the stone and the glass. Even in a small, modern cathedral such as this—a very modest building remote in time and space from its European origins—you can still see the original idea at work. The play of the sunlight on the walls and the spires, the surfaces, is really quite beautiful and mysterious to watch. The way the sunlight vivifies the stone. The way the light makes the stone seem weightless. Peculiarly, though, a cathedral is a building made to be seen from the inside, not the outside. It is like a light box. The windows are dull and dark from the outside. They light up when seen from within. This is especially true of cathedrals that boast rose windows. The great rose window of Chartres is dull and dark from outside, even when the sun strikes it. But inside, it is illumined, miraculously illumined. A gothic cathedral, by tradition, is oriented to the Sun. It is a solar building. This building is oriented. By extension of the Platonic doctrine of light it is a building set up to signal certain things about the movements of the Sun. The symbolism of the cathedral is about the movements of the Sun. The tall singularity of the spire is reiterated on the horizontal by the long aisle of the nave, another development of the gothic style not present in the Romanesque order that came prior to it. The nave extends from West to East. But why do you enter a cathedral in the West and why is the altar situated in the east? Since Constantine in the third century, Christ has been identified with the Sun, *Sol Invictus*, the triumphant Sun. Christ triumphs over death as the sun triumphs over the depletion of winter in its annual course. The sun sets in the west. The west is symbolically death, the end, the descent into darkness. When a believer enters a cathedral from the West he or she does so as a dead soul in need of salvation. The nave is the journey. The altar is the destination. On the equinox when it seems, as it were, that the whole creation forms a Cross of equal portions of night and day, the sun rises directly above the altar in an oriented gothic cathedral. Night. Day. East. West. Light. Dark. Up. Down. Life. Death. Radiance. Solidity. Glass. Stone. A cathedral, in both theory and practice, is about these dynamic polarities. A cathedral is shaped like a cross. A symbol with fairly obvious Christian meanings. But note how the cross has been laid upon the Earth and how it marks the cross of north, south, east and west. The word *geometry* means to measure the Earth. The medieval Christian—much closer to nature than we—was profoundly impressed by the fact that the Sun inscribes a cross upon the heavens in its annual movements. It moves east to west every day, but from north to south and back again through the passage of the seasons. This is the medieval cosmology reflected in the cathedral. As I said at the outset, more than any other building, a cathedral is the product of a specific world-view and a specific cosmology, and it is a world-view and a cosmology to

which the modern world no longer subscribes. But a cathedral—
even a modern cathedral in far off Bendigo in the southern
hemisphere—still works as a symbolic building—it still works in
terms of that cosmology—in the same way that a compass will
work in Bendigo just as well as in Paris. It is because it is a cosmological building, a symbol defined by the sun, fixed to the four
directions as if it is part of Creation itself, that cathedrals can be
built throughout the New World, anywhere—except perhaps in
the artic circle—and still work as living, dynamic statements of a
profoundly spiritual world-view. Because a cathedral doesn't
merely commemorate some aspect of the Middle Ages. You
don't build a new cathedral, a modern cathedral, for sentimental
reasons. As I said, the Church hasn't preserved and perpetuated
the gothic style because there are no architects about with fresh
ideas. Modern architecture—like your skyscrapers—is functional, and it may be heroic, breath-taking, awe-inspiring. It may
be witty, quirky—I'm not a critic of modern architecture per
se—but the gothic cathedral is a unique creation, a unique spiritual achievement, not merely an architectural achievement. This
is the important thing. When a community builds a new cathedral in the gothic style—however augmented—it is making a
great commitment. If nothing else, a cathedral takes a lot of up-
keep and many communities find their cathedrals a burden to
maintain. A simpler, more modest, more functional building
would make more practical sense. But a cathedral is not just a
building. Its an attitude of prayer, a prayer in stone and glass. It is
not only a place of prayer, it is a prayer itself, as it were. And I
mean this in more than the common sentimental sense. The
Christian gesture of making prayer—putting the hands together
pointed upwards—was not used by early Christians. Jesus didn't
pray in that manner. The apostles didn't pray in that manner.
That manner of praying—with the hands together and pointing
upwards—is a medieval invention from exactly the time that saw
the rise of the gothic cathedrals. In fact, that gesture of prayer
has an exact correlation to the spire of a cathedral—a symbol
of aspiration. The hands gather together and the fingers point
heavenwards like the spire. Christians began to pray in this way

to participate, as it were, in the communal aspiration heavenward—towards a spiritual life—symbolized so singularly by the
spire of a cathedral. When Christians pray in that manner they
make their hands into little cathedrals as it were. A cathedral is
not just a place of prayer. That is not the only sense in which it is
a living sanctuary. The interplay of stone and glass in the building. The orientation of the building. The play of the sun on the
outside of the building. The changing moods of light and
shadow inside. The Western portal. In the Gospel of John Christ
said, 'I am the door'. In a gothic cathedral even the doors are a
symbol, something that signals a spiritual truth, something far
beyond function.

As you drive down to Melbourne you come over one of
the hills and catch the first glimpse of the city in the distance,
the huge glass and steel towers, the skyscrapers. They look from
the distance like blocks just stuck upright upon the coastal plain.
They are an emblem of European occupation. The transplantation of European civilization from one side of the world to
another, to a new land. But they are an emblem of European
power. Might. Our economic prosperity. Our industrial
strength. Our exploitation of resources. Our technology. The
health of our markets. When you approach Bendigo and first see
the cathedral, it also speaks of a transplantation from Europe to
Australia, but it is the distinct spiritual genius of European Christendom that has been transplanted here. A remarkable spiritual
realization that found expression in architecture in the 12th and
13th C. in particular. It represents an extraordinary crystallization
of Christian ideas, Christian meditation upon existence. Sure
enough it contains diverse ideas. Ideas from Plato, then naturalized by Christian Platonists. It contains Celtic elements. It contains Islamic elements, cultural spoils from the Crusades—the
pointed arch. But it all came together in one thing as an expression of a medieval Christian impulse to praise God the creator
and to celebrate the way in which he manifests His creation
through the mystery of Light. It celebrates the way that God's
creation is inscribed with four directions, marked by the passage
of the Sun, that resplendent symbol of the Risen Christ.

A cathedral is a book. It's a history book. A cathedral is a message from the distant past. It invites us, as I said earlier, to think again about the Middle Ages, about the Christian Middle Ages, about a great era of Christian faith. But it is not a museum. It is a living sanctuary. But it is not merely, in that, a place of prayer. Its life is in being what it is. Other buildings are not like this. Hindu Temples are not like this. Jewish synagogues are not like this. Islamic mosques are not like this. Other Churches are not like this. A Buddhist stupa is not like this. There are other sacred orders of architecture in various religious traditions but a cathedral is an entirely unique idea: a building that actualizes the sacred order of Creation in *axial* terms. Think of the spire and the vertical axis. Think of the nave and the horizontal axis. The elongation of both. And this, of course, is nothing less than the symbolism of the Cross itself, the pre-eminent Christian symbol.

The difficult thing here is only that a building, by definition, is three dimensional. If you want to understand how the symbolism of a gothic cathedral works, simply extend the symbolism of the cross to three dimensions.

Think of the floor plan of this building. It's in the form of a cross. The Latin Cross, as it is called. The Greek Cross is equal armed. The Latin cross has an elongated vertical axis. The Churches of the Eastern orthodox are not like this either. Not like a European cathedral. They don't have the elongated nave. They are rectilinear, more like the Romanesque style. Instead of the spire, a dome. Instead of something that points to the sky, something that shapes the bowl of the sky. Eastern liturgical forms tend to be circular. Think of the floor plan of this building. The Cruciform shape of the Latin Cross typical of the gothic cathedrals. Aligned to the directions inscribed by the movements of the Sun. This cross, of course, any cross, symbolizes first and foremost the passion of Christ. The Cross sketches the shape, as it were, of the body of Christ, hung, tortured on the cross. The point at which the two axes meet in the building is the point from which the spire rises. I've already talked about the spirit and how it marks Bendigo's landscape today. Telecommunications dishes and towers are encroaching, but the spire of

Sacred Heart remains the conspicuous object on the horizon. In the Middle Ages they built spires of astounding height, pushing their materials as far as they would go. They pushed their spires as high as possible, until the stresses were too great and they fell down. Why were they so driven to push the spires higher and higher? I want to conclude by pointing out that if we superimpose the body of Christ upon the three dimensional symbolism of the cross, the point at which the two axes meet represents the heart of Christ. The spire represents the aspiration to Heaven—symbolized as the sky—but on another dimension it represents the aspiration to penetrate deeper the heart of Christ. Up is in. Higher and higher means deeper and deeper. The cathedral builders were driven to build spires higher and higher because, at that time, at that remarkable moment in Christian history, architecture was the art by which they were expressing an urge to contemplate deeper and deeper the profundities of the Christian faith. I think these are relevant considerations—albeit from a distant era—in a cathedral named Sacred Heart.